Second Edition
LANDSCAPING
Jack E. Ingels

For information, address Delmar Publishers Inc.
2 Computer Drive-West, Box 15-015
Albany, New York 12212

10 9 8 7 6 5 4 3

ISBN: 0-8273-2157-0

LIBRARY OF CONGRESS CATALOG CARD NUMBER: 82-74320

Printed in the United States of America
Published simultaneously in Canada
by Nelson Canada,
A Division of International Thomson Limited

CONSULTING EDITOR — H. EDWARD REILEY

Cover: Overview of the new Conservatory, formerly the Azalea House, Longwood Gardens, Kennett Square, Pennsylvania. Photo by James W. Boodley, Ithaca, New York.

PREFACE

In the past, landscaping was commonly regarded in one of two ways: as a luxury for the wealthy, or as a cosmetic for masking mediocre architecture. In its purest and most modern sense, however, landscaping represents a major defense against monotonous building styles, sprawling, unplanned suburban neighborhoods, inner city decay, and destruction of land through misuse.

The individual pursuing a landscaping career is primarily a service person whose major goal is to satisfy the needs of those who use and enjoy landscapes. To determine exactly what these needs are and how they can be satisfied without harming the lives of others or the environment is the challenge of the trained landscaper.

As an introductory text, *Landscaping: Principles and Practices* fills an educational need for those who wish to enter the landscaping field as trained professionals. By following the text, students progress from the basic principles of landscape design, installation, and maintenance to more specific topics such as choosing enrichment items for the landscape and developing cost estimates. Students learn exactly what constitutes a well-balanced, attractive landscape in harmony with its surroundings, and how such a landscape is developed.

The approach of the text encourages application of specific landscaping principles after those principles have been mastered in the classroom. Numerous photographs and drawings are inter-spersed throughout to assure understanding of content and for ease of reading. Extensive review material reinforces the learning process. Suggested Activities and Practice Exercises help to familiarize students with practices and problems they will encounter as beginning landscapers. A Glossary lists definitions of terms used frequently in the landscaping field.

An extensive Instructor's Guide is available. Besides answers to review questions, the guide includes additional Suggested Activities and a section of Class Projects. Also included are a pretest and final test, so that individual comprehension of landscaping principles may be gauged before and after the course of study.

The author of *Landscaping: Principles and Practices,* the first text in Delmar's Agriculture Series, is Jack E. Ingels. Mr. Ingels is the former Chairman of the Department of Plant Science at the State University of New York Agricultural and Technical College at Cobleskill in Cobleskill, New York. He is responsible for the campus landscaping program, in which students design, install, and maintain landscaping features of school grounds. Mr. Ingels holds a Bachelor of Science degree in Agriculture from Purdue University in West Lafayette, Indiana. His postgraduate work was done at Rutgers University, New Brunswick, New Jersey, where he received a Master of Science degree with concentration in Ornamental Horticulture, Plant Pathology, and Plant Physiology.

CONTENTS

section 1

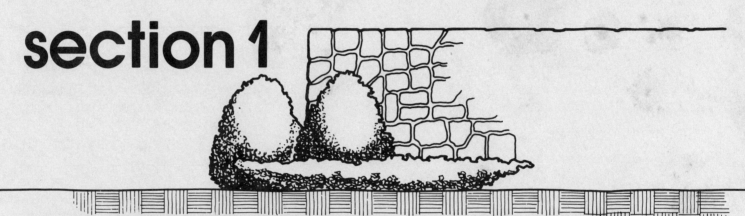

THE SCOPE OF THE LANDSCAPE BUSINESS

 unit 1

LANDSCAPE DESIGNING

OBJECTIVES

After studying this unit, the student will be able to

- explain the need for landscape designers.
- list landscape professions in which designing is all or part of the job.
- identify the tools used in landscape designing.

The term *landscaping* has many different meanings to different people. Some picture a business that grows and installs plants. Others see landscapers sitting before large drawing boards planning beautiful gardens. Still others see landscapers as those who mow lawns, prune shrubs, build patios, and erect fences. These and other tasks are a part of landscaping. For learning purposes, landscaping will be presented in three categories: design, installation, and maintenance.

THE NEED FOR PROFESSIONAL DESIGNERS

In theory, if human beings had never altered the natural world, there would be no need for designers. Nature is such a perfect designer that, left unchanged by human beings, the earth's beauty and natural system of operation would never have required improvement. This idea, however, is not realistic in the modern world. Centuries ago, people developed life-styles which set them permanently apart from the natural world. They began to grow their own food, no longer accepting what was provided naturally. They grouped themselves into living units which are our neighborhoods, cities, and suburbs. Such modern inventions as firearms, automobiles, highways, airplanes, and factories illustrate that we no longer live in a natural environment. Our population is so large that the greatest influence on our day-to-day life is not nature, but other people. The activities of each person influence the lives and activities of many others. Our activities also influence the

1

natural world, even if the activity is limited to the immediate surroundings of the home landscape.

The interrelationship that exists between the individual and the environment is complicated by the individual's desire to assert his or her own personality on the landscape. Evidence of this is seen in our business districts where every sign seems to be bigger

Fig. 1-1 This historic garden at Blenheim Palace in England is styled after the sixteenth-century Italian Renaissance gardens.

and brighter than the others around it. It is seen in our unzoned neighborhoods where hamburger stands spring up next to churches and private homes. It is also seen in our home landscapes where every house in the neighborhood may exhibit a different landscape style, attesting to the owners' particular preferences.

Our civilized world is no longer natural and no longer in a state of ecological balance. Every time an individual digs, plants, paves, or in some other way changes the land, the landscape is altered somewhat for the rest of the population. It is unrealistic to believe that all of the daily alterations people make to the land are wise and beneficial. For this reason, professional planners and designers must take a large part of the responsibility for directing society's use of land.

Good quality planning and designing are not new ideas — they have been with us for several centuries. The merits and benefits of planning and good design were understood and practiced by the wealthy and aristocratic for hundreds of years. As figures 1-1 and 1-2 illustrate, wealthy classes in the sixteenth, seventeenth, and eighteenth centuries made great changes in the landscape to create attractive outdoor spaces for their personal pleasure. These gardens were usually many acres in area and were often measures of the owners' influence and position. Certain ancient cities in Egypt, Greece, Spain, Britain, and Italy also display evidence of community planning which was sensitive to the local environment and the needs of the average city dwellers, not just the wealthy and aristocratic.

Only in the latter half of the twentieth century has the need for good planning and design been recognized at all levels of American society, including the very important middle class. Citizen involvement in community development is growing. Zoning boards of community residents are beginning to direct the growth of their cities and towns. Strict regulation of billboard placement and sign sizes is giving a new look to many business districts. Big city skyscrapers are now required to observe certain spacing regulations that allow for the development of pedestrian plazas at their bases, figure 1-3. Many towns are developing shopping malls where streets once were as one means of renewing midtown business districts, figure 1-4.

There is an increasing public awareness of the value of professional designing in the development of home landscapes. Amateur gardening gains in popularity each year and the sales of grass seed, fertilizers, trees and shrubs, and tools continue to climb. There remains a need for landscape professionals willing to direct their attention to the creation of planned landscapes for homes, commercial establishments, and other small-scale properties.

THE DESIGN PROFESSIONS

The Landscape Architect

The *landscape architect* is the professional most closely identified with landscape planning and design. In recent years, the educational requirements of the landscape architect have become much more defined and restricted. Many states now require the completion of a four- or five-year course of college study in the profession.

Fig. 1-2 The formality of this garden, including the pruning of the hedge to resemble an archway, reflects a style which was popular in past centuries throughout Europe.

Fig. 1-3 This outdoor area creates a place for relaxation amid the urban environment. This type of planning is fairly recent in America.

The course of study usually must be accredited by the American Society of Landscape Architects and leads to a B.L.A. and/or M.L.A. degree (Bachelor of Landscape Architecture or Master of Landscape Architecture). In addition, some states require a period of apprenticeship in the office of a practicing professional and/or the passing of a state licensing examination before allowing an individual to practice as a landscape architect. All of these requirements are established to safeguard the reputation of the landscape architecture profession and to protect the clients who pay for its services.

The landscape architect may be involved at all levels in the development of a landscape. It is the prime concern of the landscape architect to identify the needs of the people being served and the potential of the site to be developed. Another major concern is gathering input from other professionals and organizations that have necessary information concerning the development of the site.

Some accredited landscape architects operate at the residential level of design and involve themselves with the installation and maintenance of these properties. However, the majority of the nation's landscape architects are involved in a different type of designing and planning. These architects work with urban planners, civil engineers, and other professionals to develop cities and towns, highways, and parks, figure 1-5.

Fig. 1-4 This was once a decaying city center. The streets have been replaced by a well-planned pedestrian mall.

Fig. 1-5 A landscape architect presents his plan for a new community park to a group of interested citizens.

Fig. 1-6 The skill of the designer is reflected in the actual plan. The carefully drawn plan is the mark of a design professional.

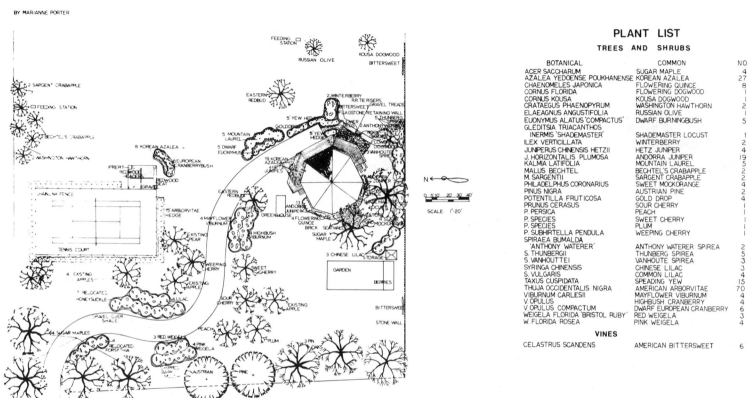

BY MARIANNE PORTER

PLANT LIST
TREES AND SHRUBS

BOTANICAL	COMMON	NO
ACER SACCHARUM	SUGAR MAPLE	4
AZALEA YEDOENSE POUKHANENSE	KOREAN AZALEA	27
CHAENOMELES JAPONICA	FLOWERING QUINCE	8
CORNUS FLORIDA	FLOWERING DOGWOOD	1
CORNUS KOUSA	KOUSA DOGWOOD	1
CRATAEGUS PHAENOPYRUM	WASHINGTON HAWTHORN	2
ELAEAGNUS ANGUSTIFOLIA	RUSSIAN OLIVE	1
EUONYMUS ALATUS 'COMPACTUS'	DWARF BURNINGBUSH	5
GLEDITSIA TRIACANTHOS INERMIS 'SHADEMASTER'	SHADEMASTER LOCUST	1
ILEX VERTICILLATA	WINTERBERRY	2
JUNIPERUS CHINENSIS HETZII	HETZ JUNIPER	4
J. HORIZONTALIS PLUMOSA	ANDORRA JUNIPER	19
KALMIA LATIFOLIA	MOUNTAIN LAUREL	5
MALUS BECHTEL	BECHTEL'S CRABAPPLE	2
M. SARGENTII	SARGENT CRABAPPLE	2
PHILADELPHUS CORONARIUS	SWEET MOCKORANGE	4
PINUS NIGRA	AUSTRIAN PINE	2
POTENTILLA FRUTICOSA	GOLD DROP	4
PRUNUS CERASUS	SOUR CHERRY	1
P. PERSICA	PEACH	1
P. SPECIES	SWEET CHERRY	1
P. SPECIES	PLUM	1
P. SUBHIRTELLA PENDULA	WEEPING CHERRY	1
SPIRAEA BUMALDA 'ANTHONY WATERER'	ANTHONY WATERER SPIREA	2
S. THUNBERGII	THUNBERG SPIREA	5
S. VANHOUTTEI	VANHOUTE SPIREA	3
SYRINGA CHINENSIS	CHINESE LILAC	3
S. VULGARIS	COMMON LILAC	4
TAXUS CUSPIDATA	SPEADING YEW	15
THUJA OCCIDENTALIS NIGRA	AMERICAN ARBORVITAE	70
VIBURNUM CARLESII	MAYFLOWER VIBURNUM	4
V. OPULUS	HIGHBUSH CRANBERRY	4
V. OPULUS COMPACTUM	DWARF EUROPEAN CRANBERRY	6
WEIGELA FLORIDA 'BRISTOL RUBY'	RED WEIGELA	3
W. FLORIDA ROSEA	PINK WEIGELA	4

VINES

CELASTRUS SCANDENS	AMERICAN BITTERSWEET	6

MR. and MRS. RON CLEEVE

Design and planting plan for:

Upper Sodom Road

Schoharie, New York

The Landscape Designer

The *landscape designer* is a professional whose academic credentials vary greatly from place to place. In some states, a landscape designer is considered to be the same as a landscape architect. As such, the designer must meet the same rigid training requirements previously described. In other states, anyone who is engaged in land design and development may be called a landscape designer. The profession descriptions which follow cover the activities of the individual who is involved in landscape design, but who does not do architectural designing.

Landscape contractors install landscapes and create designs. They are especially active in residential and commercial planning, figure 1-6. While landscape contractors often work with landscape architects on large projects, they usually do their own designing on smaller projects.

The profession of the *landscape contractor* is less regulated by governmental or trade organizations than the landscape architect. Thus, the requirements for this job vary greatly. Formal education may range from no college training to completion of degrees at several levels. The designer working with or as a landscape contractor usually creates graphic planting plans to (a) sell the client on the idea and (b) provide details for installation of the landscape. Designing at this level requires a knowledge of local life-styles, plant materials, soil and climate characteristics, and sources of material supply. Academic preparation in landscape horticulture at a two- or four-year college can be valuable to a designer in this field. Equally important is practical experience in the installation and maintenance of landscapes. Techniques observed in developed landscapes can help designers in their own planning of future gardens.

Landscape nurserymen and women use designing as a way to convince their customers to buy plants. Their profit is made from the plants and their installation charge rather than from the design. They therefore do not have to be as concerned with the graphic appearance of the landscape plan as the landscape architect or contractor. Landscape nurserymen and women use the landscape design as a sales tool. They are often least skilled in garden design, since the art of design is not their major area of concentration.

THE TOOLS OF THE DESIGNER

The creation of a professional landscape plan requires time, training, and talent. The design combines the creativity of the designer with data about the site and the needs and desires of the client. To organize all of this information into a form which others can see and understand requires techniques of graphic art.

Unlike a fine art, graphic art relies heavily on drawing tools. It is the skillful use of these tools which creates the sharp, crisp plans of the professional designer. Figure 1-7 illustrates the basic drawing tools used by the landscape designer. While many of

Fig. 1-7 Tools of the landscape designer placed on a large drawing table: (A) protractor, (B) circle template, (C) drawing pencil, (D) lead holder, (E) scale, (F) drafting tape, (G) erasers, (H) T-square, (I) compass, (J) triangles, (K) drafting powder, (L) drafting brush, (M) French curve, and (N) lead sharpener.

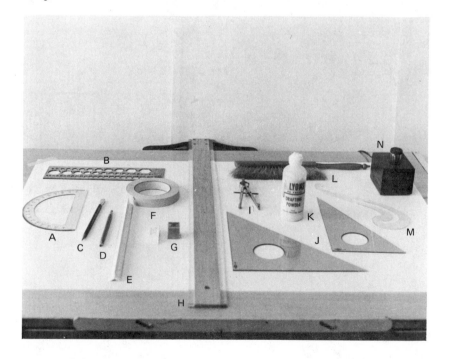

them are available in more complicated forms, their functions remain the same.

Unit 4 explains the use of the designer's instruments. At this time, the student should be able to recognize and label the design tools required by all those entering the landscape professions:

- drawing board
- T-square
- triangles
- drawing pencils
- erasers and erasure shield
- drafting tape
- scale
- protractor
- compass
- circle template

ACHIEVEMENT REVIEW

A. Write a short essay on the reasons why professionally trained landscape designers are needed today. Include: (a) the relationship of the individual with nature; (b) the interrelationship among people within the environment; (c) examples of good planning and design throughout the country; and (d) examples of good and poor planning in your local community.

B. Match each job on the left with its characteristic(s) on the right.

a. Landscape architects
b. Landscape contractors
c. Landscape nurserymen and women

1. Their main interest is selling plants.
2. This profession usually requires the greatest number of years of schooling.
3. They construct landscapes as well as design them.
4. They are often involved in planning large urban projects.
5. They usually are not too concerned about the appearance of a design; it is mainly a selling tool for them.

C. Identify each of the following drawing tools.

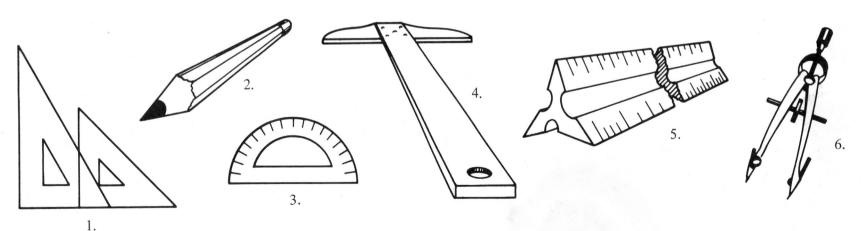

1.
2.
3.
4.
5.
6.

 # unit 2

LANDSCAPE INSTALLATION

OBJECTIVES

After studying this unit, the student will be able to

- describe the job of the individual who installs landscapes.
- explain the relationship between the design and installation of landscapes.

Once a landscape design has been completed and received approval from the client, it must be installed. *Landscape installation* is the actual construction of the landscape. The process of installation lifts the design from paper and brings it to reality. It is the landscape contractor who is most often involved in large landscape installations. Even if the job is the responsibility of a landscape architect, the landscape contractor is often subcontracted to do the actual installation. In cases which require only the use of plant materials, the landscape nurseryman or woman may also do some installation.

LANDSCAPING AS A BUILDING TRADE

Those who earn a living by installing landscapes are similar in some ways to persons who work in the building trades. Many elements of landscaping require definite knowledge of specific construction techniques to bring them into being, figures 2-1 and 2-2. For example, imagine that a design calls for a poured concrete patio and a brick wall. What associated questions must be answered by the installer before construction can begin?

Poured Concrete Patio

- How is concrete made?
- How much concrete is needed?
- How thick must the concrete be to avoid cracking without being too thick?

- How is the wooden form to be constructed?

- Is some wood better than others for construction of the form?

- How much time is required for concrete to harden?

- Should methods that force concrete to harden more quickly or slowly than is natural be employed?

- Should the patio surface be smooth or rough, and how is either accomplished?

- How can the patio be constructed so that water drains off, yet it still appears level?

- Is concrete poured directly onto the soil or does it require a base?

Brick Wall

- Are all bricks the same size?

- Are all bricks the same strength?

- Should hardness of the brick be considered?

- How high can a wall which has a thickness of a single brick be built? At what height is a second thickness needed?

- Does turning a corner weaken or strengthen a brick wall?

- What are the ingredients of mortar?

- Should brick be moist or dry during installation?

- Is it preferable for mortar to dry quickly or slowly?

Fig. 2-1 **Surveying skills are often needed by professionals who install landscapes.**

Fig. 2-2 **The landscape contractor must know how to work with many construction materials. Here, a stone wall is under construction.**

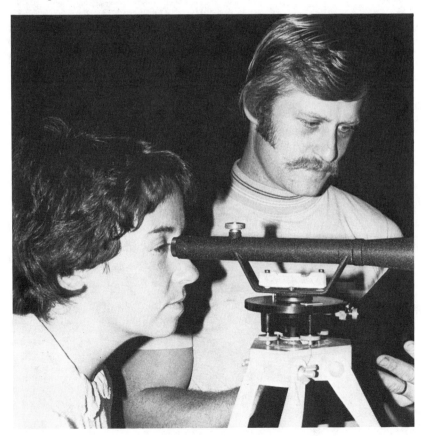

- How much space should be left between bricks?
- How are bricks cut?

These two examples illustrate some of the knowledge and skills of the construction trade which might be required by the landscape contractor. Other areas could be included in the list, such as plumbing, lighting, wiring, woodworking, and use of heavy equipment and hand tools.

However, when plants are included as part of the installation job, the landscape contractor has a unique job among persons in the building trades. Being living things, plants require much more care during installation to prevent injury than do building materials such as bricks and concrete, figures 2-3 and 2-4. It is important that the person installing landscapes know how to prepare a planting site that assures the success of the trees, shrubs, grass, flowers, vines, and ground covers.

Fig. 2-3 Many plants are installed by hand. The landscaper must know how to prepare the planting site to assure a successful transplant.

Fig. 2-4 Large trees may require special handling to assure successful relocation. Here, a tree spade is used to prepare the ground for planting of a large maple.

Fig. 2-5 A large tree is felled to prepare the site for landscaping. The landscaper wears protective clothing and sound mufflers for safeguarding her hearing.

FIRST THE DESIGN, THEN INSTALLATION

The landscape installer must work closely with the landscape's designer if the best results are to be obtained. The designer should provide the installer with a copy of the landscape plan on durable paper or plastic film. The material used should allow the plan to withstand repeated folding and unfolding. This *working drawing* becomes the instruction sheet for the landscape's installation. From it, the installer is able to determine such things as the exact placement of each plant, the exact width and length of each planting bed, and the exact dimensions of each patio, wall, and fence. Without the plan, the installation can result in a collection of overcrowded, incorrectly placed plants creating a maintenance nightmare after a few seasons of growth.

ACHIEVEMENT REVIEW

A. Provide a brief explanation of each of the following.

1. What makes a landscape contractor different from others who work in the construction field?

2. What is a working drawing? Who provides it? Who uses it?

3. What is the value of a landscape plan to the landscape installer?

B. Select the best answer from the choices offered to complete each statement.

1. The landscape installer must work closely with the landscape's _____ if the best results are to be obtained.

    a. contractor   b. designer   c. gardener   d. maintenance crew

2. Those who earn a living by installing landscapes are similar to those who work in the _____ trades.

    a. maintenance   b. building   c. artistic   d. nonskilled

3. A landscape contractor is often subcontracted for installation work by a landscape _____.

    a. foreman   b. supervisor   c. gardener   d. architect

4. To assure the success of the plant materials, a good landscape installer knows how to prepare a (an) _____.

    a. planting site   b. bid   c. estimate   d. bare-root plant

SUGGESTED ACTIVITIES

1. Select several landscape features either near the classroom or from magazines. These items might be walls, fences, fountains, pools, or light fixtures. Hold sessions in which the entire class or smaller groups of students analyze the many questions that must be answered during construction of those particular features.

2. Make a list of the construction skills that you currently possess. Pair off with a student who has similar skills (such as bricklaying or carpentry). With your partner, report to the class on the special tools and materials required by those skills.

 **unit 3**

LANDSCAPE MAINTENANCE

OBJECTIVES

After studying this unit, the student will be able to

- list the types of jobs done in landscape maintenance.
- list the types of positions which involve landscape maintenance.
- identify tools commonly used in landscaping.
- explain the relationship between the designer and the maintainer of landscapes.

Landscape maintenance is the care and upkeep of the landscape after its installation. Since landscapes are *dynamic* (constantly growing and changing), they alter their appearance and size each season, thereby requiring different types of maintenance at different times of the year.

The following is a partial listing of the jobs necessary for year-round maintenance of a landscape.

- mowing of the lawn
- pruning of trees and shrubs
- application of fertilizer to lawn and plantings
- weed control in lawn and plantings
- spraying and/or dusting for insect and disease control
- planting and care of flower beds and borders
- replacement of dead plants
- painting or staining of fences and outdoor furnishings
- repairing of walls and paved surfaces
- cleaning of fountain and pool basins
- irrigation of lawn
- cultivation of soil around trees and shrubs
- replacement of mulches
- removal of lawn thatch

- rolling and reseeding of lawn
- raking of leaves in fall
- winterization of trees and shrubs
- snow removal
- preventive maintenance on equipment

Several of these jobs can be done by the homeowner or by an untrained employee. Most, however, require specialized equipment and training. This places complete and top quality landscape maintenance beyond the capability of the amateur. Because of the increasing demand for professional grounds maintenance, certain landscapers direct most or all of their time to this work. Full-time landscape maintenance persons are usually called *grounds keepers* or *gardeners*.

It might be expected that all landscape architects, contractors, and nurserymen and women offer landscape maintenance services to their clients. Certainly some of them do. However, the majority are not involved in grounds maintenance because they do not want to purchase the necessary specialized equipment. Much of this equipment is not usable in other types of landscape work. Also, the profit per hour of landscape maintenance is often not as great as that for design and installation.

In most areas of the country today, the public demand for professional landscape maintenance exceeds the number of properly trained landscapers willing to provide those services. This gap between supply and demand is being filled by an assortment of skilled and unskilled amateurs working part time or full time at the business. The limited services usually offered by the amateur include lawn mowing, pruning, weeding, and yard cleanup.

The public has been led to believe that landscaping, as represented by the part-time amateur, is a low-cost service which requires no formal training or knowledge. The amateur, however, simply cannot provide the full line of services required of the professional. When clients encounter a fully qualified landscape maintenance professional, they are often surprised by the scope and cost of proper landscape care.

For a person interested in beginning a career in the landscape business with a relatively small cash investment, landscape maintenance is a good choice. There is no need for land, greenhouses, or extensive building space. Most of the money initially invested is placed toward the purchase of proper equipment necessary to do a high quality job. Materials such as fertilizer, grass seed, and mulches can be bought from a supplier and their cost charged back to the customer. The charge for services is the major source of income for the independent landscape grounds keeper.

In addition to self-employment, there are other ways to become involved in landscape maintenance. Most large parks, businesses, and institutions maintain their own grounds crews. This often reflects the shortage of qualified private maintenance enterprises in their areas. Employment as a grounds supervisor for a large public or commercial landscape can be very rewarding. It provides the opportunity to see the landscape grow and mature through several seasons. A sense of accomplishment and pride develops that is missed by the grounds keeper who moves regularly from job to job.

There are also a limited number of jobs available as gardeners for private estates. Although the career of resident gardener remains strong in various European countries, it is on the decline in America.

TOOLS OF THE TRADE

High quality landscape maintenance requires a wide variety of tools, some of which serve many purposes. Others are very specialized. Many items, such as trucks, lawn mowers, wheelbarrows, and hoses, are in common use. Following are some of the more specialized hand tools. Students should learn to recognize each tool and know its proper name and function.

Tool and Name	Function	Tool and Name	Function
Grass Shears	Used to trim grass along walks, roadways, the edge of planting beds, and around trees, posts, etc.	**Hedge Shears**	Prunes shrubs grown closely spaced as hedges. These shears are only used on young, tender new growth.
Pruning Shears	Used to trim tree and shrub twigs up to one-half inch in diameter.	**Pruning Saw**	Removes any tree or shrub part which cannot be easily cut with the lopping shears. Usually parts are an inch or more in diameter.
Lopping Shears	Used to trim tree and shrub twigs from one inch to one and one-half inches in diameter.	**Crosscut Saw**	Removes large limbs and small trees. The saw has additional general uses.

Tool and Name	Function	Tool and Name	Function
Grass Hook	For reducing the height of overgrown grass areas. It requires the user to bend over.	**Spades**	Obvious general uses in digging. Spades have flatter shapes than shovels. They penetrate the soil more easily but have less scooping capability.
Grass Whip	For reducing grass height without bending over. (Once reduced in height, a lawn mower can be used on the grass.)	**Shovel**	Used for cleaning loose soil from planting holes and other scooping uses. A shovel has sides that a spade does not have.
Spading Fork	Used for turning over the soil when it is not too hard or compacted. Also used for lifting bulbs in the fall.	**Spading Shovel**	A combination tool having uses similar to both spades and shovel. It can be used for digging as well as scooping.

Tool and Name	Function	Tool and Name	Function
Scoop	Good for moving loose materials such as crushed stone, peat moss, soil, etc. Scoops have high sides. They are not used for digging.	**Weed Cutter**	Removes annual weeds by cutting them off at ground level. Not very effective against biennial and perennial weeds.
Manure Fork	The best tool for moving coarse, lightweight materials such as straw, wood chips, etc.	**Toothed Rakes**	Used for heavy duty raking which requires a strong tool. Commonly used in preparation of lawn seed beds and cultivation of planted beds.
Single-Bit and Double-Bit Axes	Obvious chopping uses. Especially useful in tree removal and for cutting up fallen timber.	**Broom Rake**	Very useful in places where a lightweight springy rake is needed. Very good for collecting debris and clippings from lawn surface.

Tool and Name	Function	Tool and Name	Function
Lawn Comb	An excellent rake for collection of leaves and coarse debris from lawn surface.	**Hand Trowel**	Used for transplanting bedding plants into flower beds, borders, and boxes.
Shrub Comb	Used for raking debris from small areas between shrubs.	**Transplanting Hoe**	Uses are similar to those of a hand trowel. It has less adaptability for other types of digging.
Bulb Planter	Used to install flowering bulbs.	**Scuffle Hoe**	Useful in weeding and cultivating in planted beds. It cuts off weeds and loosens surface soil.

Tool and Name	Function	Tool and Name	Function
Push Hoe	Similar to a scuffle hoe. It is good for rooting out weeds.	**Post Hole Digger**	Prepares holes for the support posts of fences.
Garden Hoe	Widely used for breaking up the soil prior to planting. It is also good for cultivating planted beds and for weed removal.	**Grading Hoe**	Loosens hard or compacted soil during preparation for planting. Has a sharpened flat end.
Hand Cultivator	Loosens the surface soil in flower beds and around shrubs.	**Cutter Mattock**	Stronger than a grade hoe. Its uses are similar. It has two flat ends.

Tool and Name	Function
Pick	Used for breaking up hard rocky soil. It has two pointed ends for gouging into the soil.
Spreader	Used for the application of fertilizer, seed, and other dry turf products.
Sprayer	Needed to apply pesticides, antitranspirants, and other chemicals in liquid form. Sprayers are available in a wide range of sizes.

THE RELATIONSHIP BETWEEN LANDSCAPE DESIGN AND LANDSCAPE MAINTENANCE

It is important that the landscape designer be concerned with the amount of care necessary to keep the landscape attractive. If designers are thoughtless in their planning, the grounds keeper may be faced with many problems. For example, some plants are known to be especially apt to attract insects and disease, requiring a great deal of costly spraying to keep them healthy looking. Other plants which are less susceptible can be used to fill the same role, thereby reducing maintenance. Certain plants have demanding soil requirements. The use of these plants can also cause extra work for the grounds keeper. The designer might specify other plants which require less attention.

Other design specifications can act to lighten or burden the job of the landscape grounds keeper. A smooth-edged, flowing bedline is easier to mow along than a fussy complex bedline, figure 3-1. Where walks intersect at 90-degree angles, a specific design can minimize wear on the lawn; this in turn requires less maintenance time, figure 3-2.

In brief, the amount of care required by a landscape can be controlled to some extent by the designer. It is the responsibility

Fig. 3-1 Landscape maintenance can be made easier by a thoughtful designer. Bedline A is fussy and complex. Mowing along this bedline would be a time-consuming process. Bedline B, more thoughtfully designed, would be easier to maintain.

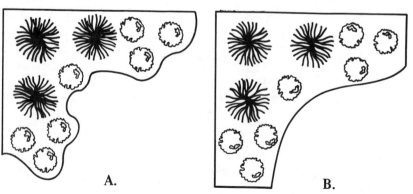

A.   B.

of the designer to know the requirements of landscape mainten- ance so that maintenance time can be kept to a minimum. If possible, grounds keepers responsible for maintenance should be given the opportunity to review landscape plans early in their development so that their suggestions for easier maintenance can be taken into consideration.

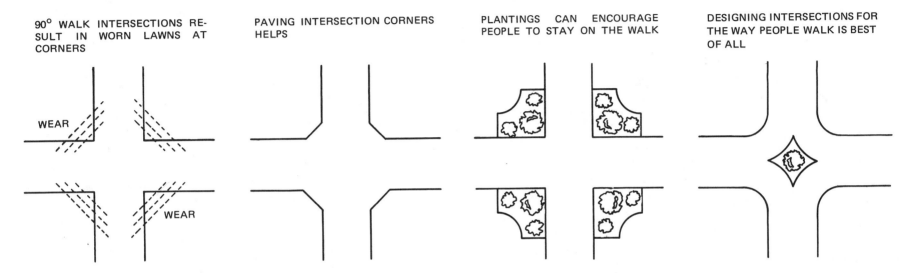

Fig. 3-2 Various design techniques used to reduce wear on lawns.

90° WALK INTERSECTIONS RE-SULT IN WORN LAWNS AT CORNERS

WEAR

WEAR

PAVING INTERSECTION CORNERS HELPS

PLANTINGS CAN ENCOURAGE PEOPLE TO STAY ON THE WALK

DESIGNING INTERSECTIONS FOR THE WAY PEOPLE WALK IS BEST OF ALL

ACHIEVEMENT REVIEW

A. Select from the list below only those jobs which are done as a part of landscape mainte-nance.

constructing patios
mowing lawns
pruning dead tree limbs
designing landscape plans
fertilizing lawns

raking leaves
preparing new planting beds
adding fresh mulch to plantings
installing swimming pools
building fences

B. Identify and spell correctly the following tools.

1.

2.

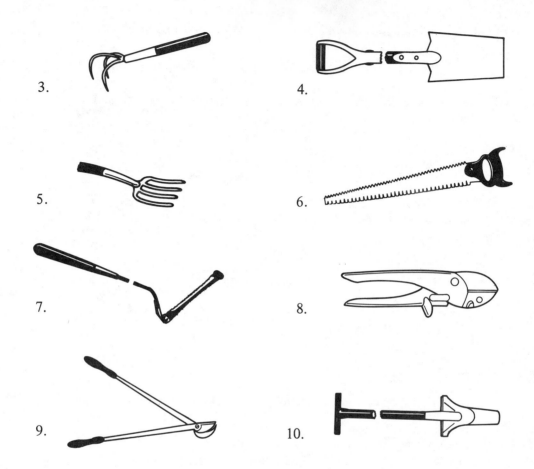

3.

4.

5.

6.

7.

8.

9.

10.

SUGGESTED ACTIVITIES

1. Do a study of maintenance requirements of your school's landscape. Select ten different features (such as shrub beds, trees, parking lots, and entrances) and rank them in order of ease or difficulty of maintenance.

2. Itemize all of the maintenance requirements necessary for the proper upkeep of each feature given in your answer to question 1.

3. Match the maintenance requirements listed in question 2 with the proper tools needed to accomplish the job.

section 2

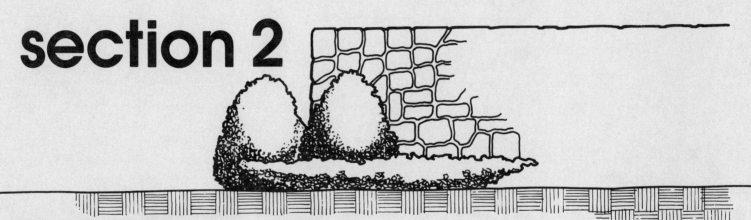

PRINCIPLES OF
LANDSCAPE DESIGNING

 # unit 4

USING DRAWING INSTRUMENTS

OBJECTIVES

After studying this unit, the student will be able to

- properly use the instruments important in landscape designing.
- measure and duplicate angles.
- measure and interpret lengths and distances to scale.

The landscape designer is concerned with the creation of the landscape plan. After an idea originates mentally, it is transferred into a form in which it relates to the total project. Still later, the designer presents plans for the landscape that the client can see and understand.

Landscape designers use several different mechanical drawing instruments to transfer their ideas from their minds onto paper. The *T-square* is a long straightedge which takes its name from its shape, figure 4-1. When used with a smooth-surfaced *drawing board*, with four 90-degree corners, the T-square can be used to draw a series of horizontal or vertical lines which are parallel, figure 4-2. It is important that the T-square be kept flush with the

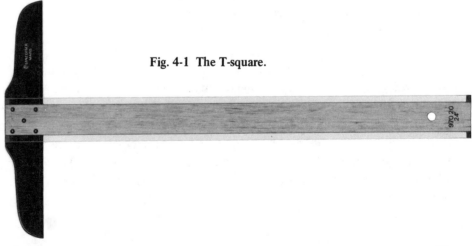

Fig. 4-1 The T-square.

Fig. 4-2 The T-square may be used to create parallel, horizontal, or vertical lines.

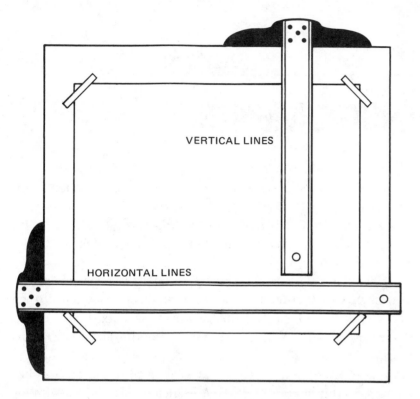

VERTICAL LINES

HORIZONTAL LINES

A *compass* is used to create circles, figure 4-3. For the landscape designer, these circles form the guidelines for creating tree and shrub symbols. (Explained in Unit 6). The metal pointed leg of the compass is placed in the center of the circle. The pencil or leaded leg of the compass creates the arc. When drawing circles, it is important to remember that the distance between the two legs of the compass should be one-half the desired diameter of the circle. For example, if a circle with a 4-inch diameter is to be the result, the compass legs are set at a spacing of 2 inches, figure 4-4.

Fig. 4-3 Compass.

edge of the drawing board at all times when parallel or perpendicular lines are being drawn. The T-square is used by the landscape designer to create lines to represent property lines, walks, roads, driveways, and fences.

The *drawing pencil,* a basic tool of the landscape designer, may be an expensive or inexpensive instrument, depending upon the quality selected. The designer may select from wooden pencils with leads of varying hardness or plastic and/or metal lead holders whose various leads are purchased separately and inserted. It is recommended that the beginning designer select inexpensive wooden drawing pencils. Both 2H and 3H leads should be obtained. The *H rating* of a pencil is a measure of its hardness. The higher the H rating, the harder the lead is. The lead should have a sharp point at all times.

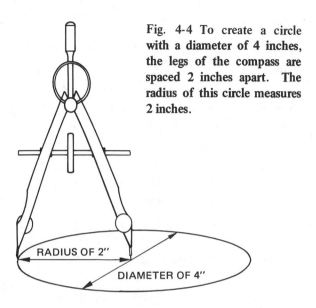

Fig. 4-4 To create a circle with a diameter of 4 inches, the legs of the compass are spaced 2 inches apart. The radius of this circle measures 2 inches.

A *protractor*, figure 4-5, measures the relationship between two joined lines. This relationship is known as an *angle;* the unit of measurement is a *degree.* To measure an angle, the center notch at the base of the protractor is placed upon the point at which the two lines join. The 0-degree mark of the protractor's baseline is aligned along the lower line being measured. The angle is determined by reading up from 0° to the point at which the second line intersects the protractor. In figure 4-6, the protractor is measuring a 25-degree angle. It is important to remember that the reading is always taken between two existing lines, starting at 0°. In this way, confusion over the protractor's double scale is avoided.

Several plastic or metal *triangles* are also used by landscape designers. The two most common triangles are those having 30°-60°-90° and 45°-45°-90° angle combinations, figure 4-7. The triangles are frequently used as straightedges by themselves, but are also used in combination with the T-square to create angles of 30, 45, 60, or 90 degrees, figure 4-8.

The most vital instrument used by landscape designers is a measuring tool known as a *scale,* figure 4-9. While the scale

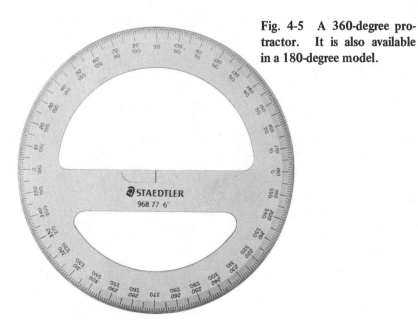

Fig. 4-5 A 360-degree protractor. It is also available in a 180-degree model.

Fig. 4-6 The protractor is placed along one line of the angle and aligned with zero (0°). The angle is read at the point of intersection of the second line.

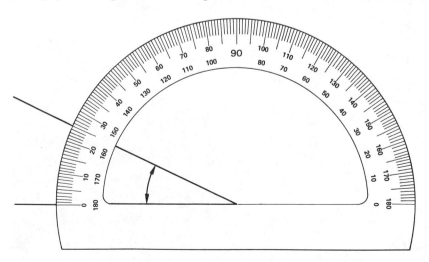

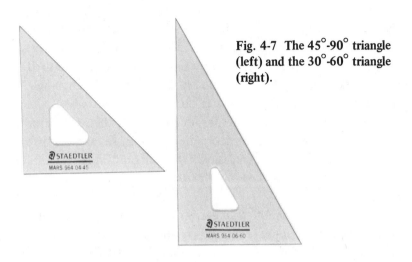

Fig. 4-7 The 45°-90° triangle (left) and the 30°-60° triangle (right).

Fig. 4-8 The triangle can be used with the T-square to create angles of 30, 45, 60, or 90 degrees.

resembles an ordinary ruler, it has many more uses than simply measuring inches. The engineer's scale, which divides the inch into units ranging from ten to sixty parts, is the most easily read scale.

The scale instrument is used by the landscape designer to represent or reproduce actual land dimensions and objects on the drawing paper at a size convenient for working. For example, if a property line is 90 feet long, it can be represented on the designer's paper as a line 9 inches long if the scale of the drawing is 1" = 10'. The same length can be represented by a line 4 1/2 inches long if the scale of the drawing is 1" = 20'. In figure 4-10, the same measurement of 90 feet is located on each of the six sides of the scale instrument. Note that each unit represents 1 linear foot regardless of the side of the instrument used, as long as the scale of the drawing corresponds to the proper side of the scale instrument.

Once its use is mastered, the scale instrument permits the designer to represent entire building lots, houses, walks, plants, and other items on paper. If the same scale is used throughout, all objects will be in the proper relationship to one another.

Fig. 4-9 The scale. The triangular shape of this instrument creates six measuring sides.

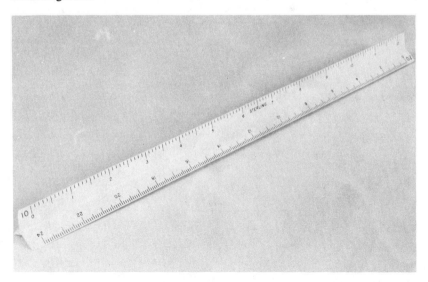

Many other drawing tools are available to the landscape designer. Most are used to keep the drawing neat and clean. Included are such items as erasers, erasure shields, drafting powder, and pens.

Beginning designers can purchase the basic instruments fairly inexpensively. Many of the items can be found in variety stores. Several of the items, such as the drawing board, T-square, and scale, can be purchased at drafting supply stores or art supply stores.

Fig. 4-10 Reading the scale. The six views show where one measurement (90 feet) appears on each side of the scale instrument. (Individual foot units are not shown in the scales on the bottom line.)

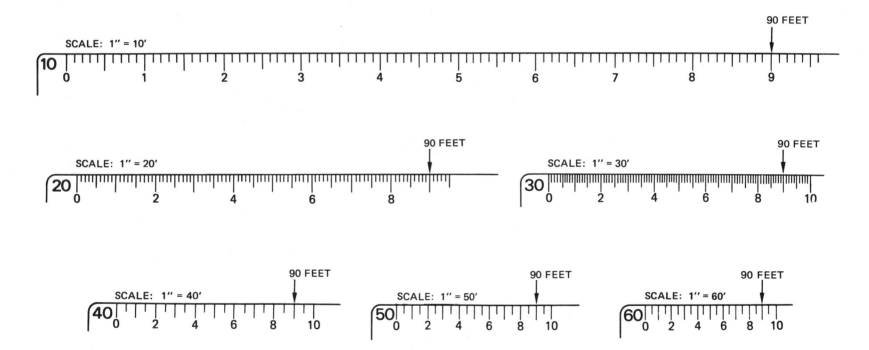

PRACTICE EXERCISES

1. Tape a piece of drawing paper onto a drawing board. Practice drawing parallel horizontal lines and vertical lines using a T-square. Use a 2H pencil first and then a 3H or 4H pencil. Which pencil marks smear most easily? Which are easiest to erase?

2. With your protractor, measure the angles in the figure below. Duplicate the angles on a separate sheet of paper.

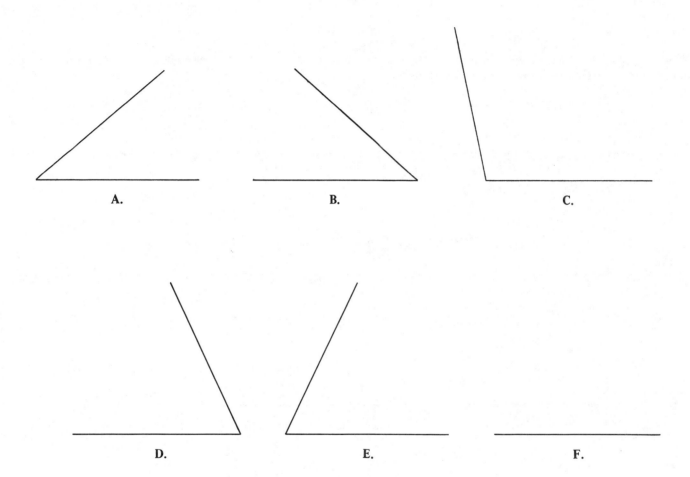

A.

B.

C.

D.

E.

F.

3. Using a scale instrument to measure and a triangle as a straightedge, draw a line 30 feet long to the scale of 1″ = 10′. Draw another 30-foot line to the scale of 1″ = 50′. Draw an 87-foot line to the scale of 1″ = 40′ and another to the scale of 1″ = 30′.

4. Use a compass and a scale instrument to create a circle with a 20-foot diameter drawn to the scale of 1″ = 10′. What is the radius setting for the compass? Draw another circle with a 20-foot diameter to the scale of 1″ = 20′.

ACHIEVEMENT REVIEW

A. Explain the function of each drawing instrument listed.

T-square
triangle
compass
protractor
scale

B. From the choices offered, select the best answer to each question.

1. A compass set for a 2-inch radius forms a circle of what diameter?

a. 2 inches   b. 4 inches   c. 1 inch   d. 6 inches

2. If the scale of the drawing is 1″ = 5′, the completed circle described in question 1 would be how many feet wide?

a. 5 feet   b. 10 feet   c. 15 feet   d. 20 feet

3. Which of the following indicates the hardest pencil lead?

a. 2H   b. 3H   c. 4H   d. 5H

4. Which two instruments combine most easily to make a 90-degree angle?

a. compass and scale   c. protractor and scale
b. T-square and triangle   d. compass and T-square

5. What is the most logical way to measure an angle when the lines are too short to intersect the protractor?

a. Do not measure it; simply estimate it.
b. Attempt to trace the angle.
c. Extend the lines of the angle until they intersect the protractor.
d. Measure it with the scale instrument.

unit 5

LETTERING

OBJECTIVES

After studying this unit, the student will be able to

- name three methods of lettering used in landscape designs.
- compare the advantages and disadvantages of each method.
- create four different styles of free-hand lettering.

THE IMPORTANCE OF LETTERING

A quick glance at a professionally prepared landscape plan reveals that it is a blend of both symbols and lettering. The symbols suggest how the proposed objects and landscape will appear. The lettering usually identifies the objects and often explains how they are used or installed. The lettering requires the same care and quality that is applied to the formation of the symbols. A plan that is used as a device for selling the landscaper's ideas to the client must be prepared with the highest graphic standards. This includes the neatest and most attractive lettering. To the client, shoddy lettering will suggest shoddy design, although that association may not be totally fair or logical.

METHODS OF LETTERING

Today's sophisticated advertising layouts fill magazines, newspapers, and television with intricately styled lettering that is colorful, and both difficult and time consuming to create. In contrast, the labeling of a landscape plan requires lettering that is

- easy to create.
- rapid to apply.
- attractive.
- compatible with the symbol styling, without overpowering it.

Three methods of lettering that fulfill these requirements are: waxed press-on letters, letters created with stencil guides, and letters created by the designer's own hand. These methods are commonly referred to, respectively, as press-ons, lettering guides, and freehand. Each method has advantages and disadvantages as compared to the others.

Waxed Press-on Letters

When a landscape plan requires a highly refined, professional lettering style, the waxed press-on letters are the quickest and easiest method to use. They are commercially manufactured in an assortment of styles and sizes, figure 5-1. The letters are mounted and sold on sheets of plastic or waxed paper, figure 5-2. They are transferred onto the drawing surface by the heat of friction created when a pencil is rubbed over the surface of the sheet, figure 5-3. To improve their adhesion, the letters are then rubbed again with the pencil through a sheet of waxed paper that usually accompanies each sheet of letters, figure 5-4.

The following are the major advantages of press-on letters:

- They are easy to apply
- They create a highly refined graphic impression
- They are easily removed, permitting the correction of mistakes

Fig. 5-1 Styles and sizes of press-on letters

35 **Folio Extra Bold**	45 **Gill Extra Bold**	61 **News Gothic Bold**	15 *Berling Italic*
36 **Folio Bold Condensed**	47 Grotesque 7	61 News Gothic Condensed	16 **Berling Bold**
36 **Franklin Gothic**	47 **GROTESQUE 9**	66 **Pump Medium**	16 Beton Medium
37 *Franklin Gothic Italic*	47 *Grotesque 9 Italic*	66 **Pump**	16 Beton Bold
37 **Franklin Gothic Cond.**	48 Grotesque 215	69 **Simplex Bold**	17 **Beton Extra Bold**
38 **FRANKLIN GOTHIC EX COND**	48 **Grotesque 216**	71 **Standard Medium**	17 Bookman Bold
39 ᗡNOƆ ƆIHTO⅁ NIⅬ⋊NA刊ꟻ	49 Helvetica Ex Light	72 **Standard Extra Bold Cond.**	18 *Bookman Bold It.*
39 Futura Light	50 Helvetica Light	75 Univers 45	20 **Carousel**
40 Futura Medium	51 Helvetica Medium	75 Univers 53	20 Caslon 540
40 *Futura Medium Italic*	52 **Helvetica Bold**	76 Univers 55	21 **Caslon Black**
41 **Futura Bold**	52 *Helvetica Light Italic*	76 Univers 57	21 **Century Schoolbook Bold**
41 ***Futura Bold Italic***	53 ***Helvetica Med. Italic***	76 Univers 59	22 Cheltenham Old Style
42 **Futura Demi Bold**	53 ***Helvetica Bold It.***	77 Univers 65	22 Cheltenham Med

Fig. 5-2 A sheet of one style of press-on letters

Fig. 5-3 Press-on letters are transferred onto the design by the heat of friction.

The following are the major disadvantages of press-on letters:

- They are expensive
- Frequently used letters run out before others are used, causing costly waste
- They crack with age

Press-on letters are commonly used for presentation drawings and for important concept drawings where the graphic appearance of the designer's work must be first-rate. They are often used both by landscape architects and landscape contractors.

Fig. 5-4 To secure them to the surface, the letters are rubbed again through a sheet of waxed paper.

Fig. 5-5 Using a lettering guide

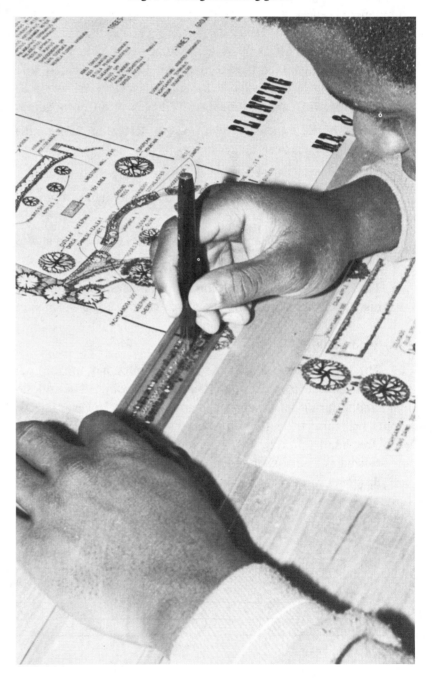

Lettering Guides

Lettering guides are templates that reproduce the same letter size and style over and over again. They may be used with a drawing pencil or lead holder, but are most often used with technical pens. If a technical pen and ink are used, a lettering guide sized to match the width of the pen's tip must be selected. The guide must also have raised edges to permit its use without smearing the ink, figure 5-5.

Lettering guides in several styles and sizes are available on the commercial market. A single landscape drawing may require letters of varied sizes. Thus, a selection of lettering guides may be needed. The drawing usually looks best if the same style of letters is used throughout. Using mixed brands of guides should be avoided, since this can result in varied styles.

The following are the major advantages of lettering guides:

- They are less expensive than press-on letters once the initial investment has been made

- There are no leftover letters; therefore, there is no waste

- They allow a designer who letters poorly by hand to produce an attractive plan

The following are the major disadvantages of lettering guides:

- They are time consuming for some people to use

- They remove the human element that makes free-hand lettering interesting; thus, the design often looks too mechanical

Free-hand Lettering

The free-hand method is the most common and most important method used for lettering a landscape drawing. The styles of free-hand lettering can be described and studied. However, no two people will form the letters in exactly the same way. As a result, good free-hand lettering has a personality and quality as unique and stylized as a person's handwriting.

Students can develop a good free-hand lettering style by carefully analyzing how the letters are constructed, and by repeatedly practicing the new lettering styles so as to replace the ones learned earlier in life.

STYLES OF FREE-HAND LETTERING

The styles of free-hand lettering used by most landscape designers are usually based upon four traditional styles:

- Basic block letters

- Distorted block letters

- Slanted block letters

- Slanted, distorted block letters

Each of these basic styles is merely a variation of the others. All four styles create single-stroke letters the width and thickness of one pen or pencil line. Because they are simple to construct, the letters may be produced quickly and with consistency of width and style. All styles require the use of horizontal guidelines and, when large in size, the support of straightedges may be needed. Generally, in common practice, the letters are created without the need for straightedges; the designer relies only upon the T-square to keep the lettering properly aligned.

Each of the free-hand styles described here creates letters that are either equal in height and width or else slightly higher than they are wide. All are based upon three horizontal guidelines. The *placement of the horizontal midline* is one determining factor in creating the styles. The *angle of the vertical slant* of each letter is the other determining factor.

Basic Block Letters

Basic block letters are the basis of all free-hand lettering styles. The horizontal midline is equidistant between the upper and lower lines. All vertical lines are drawn at a 90-degree angle to the horizontal lines. Figure 5-6 illustrates the complete alphabet lettered

Fig. 5-6 The alphabet in basic block style

in the basic block style. Note that the midlines of the letters A, B, E, F, G, H, P, R, and Y touch the center guideline. Certain letters, such as M, O, Q, and W are as wide as they are tall. Others, except the I, are about two thirds to three quarters as wide as they are tall. To keep the letters properly vertical, some designers strike 90-degree reference lines at random intervals along the line being lettered as a guide for their eyes.

Figure 5-7 shows a line of letters being developed. Note the placement of the T-square for construction of the horizontal guide-lines and the use of the 90-degree triangle to provide the vertical reference lines. Still, the actual letters are created freehand.

If a technical pen is used for the lettering, the line width will remain consistent, as it should. If a pencil or lead holder is used, the designer must frequently repoint the lead in a sharpener or by rolling the point across a sandpaper pad, figure 5-8.

Another aspect of free-hand lettering that applies to the basic block style, as well as the three variations, concerns serifs. *Serifs* are decorative strokes attached to letters to create a more ornate

Fig. 5-7 Developing basic block letters using the T-square and 90-degree triangle

Fig. 5-8 Sharpening the point of a drawing pencil on a sandpaper pad

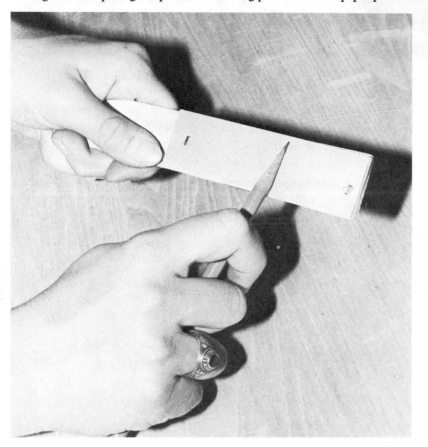

style. Serif-style type, which is common to most typewriters and newspapers, is used in this text. It is so common that many students apply serifs to certain letters, notable the I and J, even when lettering in a nonserif style. Since landscape free-hand lettering is a nonserif style, there are no serifs on the I and J or any other letters.

Distorted Block Letters

One variation of the basic block style is created by either raising or lowering the horizontal midline from its center position. Letters which touched the midline at center before now touch a raised or lowered midline. The result is a distortion of the basic block lettering, a style preferred by many designers. Figures 5-9 and 5-10 show the alphabet in distorted lettering styles.

Slanted Block Letters

In the basic block style the three horizontal guidelines are equidistant and the vertical reference lines are at a 90-degree angle in relation to the horizontal lines. Another lettering variation can be created by angling the vertical reference lines and the vertical sides of the letters at 60 degrees instead of 90 degrees. The result is a slanted block style of lettering, figure 5-11. Some designers prefer this style, especially if their own handwriting tends to be more slanted than vertical.

Fig. 5-9 Distorted block alphabet with midline raised

Fig. 5-10 Distorted block alphabet with midline lowered

Fig. 5-11 Slanted block alphabet

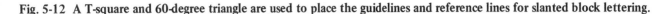

Fig. 5-12 A T-square and 60-degree triangle are used to place the guidelines and reference lines for slanted block lettering.

In figure 5-12, the letters are shown being constructed with a T-square and a 60-degree triangle in place to provide guidelines and reference lines. A student attempting slanted block lettering is likely, at first, to find it difficult to maintain a consistent 60-degree slant. Mastering this style takes time and practice. The 60-degree reference lines can be of great help.

Slanted, Distorted Block Letters

The final variation of the basic block style is actually a combination of the variations described earlier. In this style, the horizontal midline is raised or lowered from the center, *and* the vertical

orientation of the letters and reference lines is 60 degrees. Figure 5-13 illustrates the alphabet lettered in the slanted, distorted block style.

DEVELOPING A LETTERING STYLE

A textbook can only offer guidance to a new designer seeking to develop a good lettering style. Each designer must practice and modify his or her style until the desired results are attained. The style that develops is likely to be a personalized version of one of the four free-hand styles previously described. It may be further modified to suit the designer's own preference and style. Lettering

Fig. 5-13 Slanted, distorted block alphabet with guidelines. Half of the alphabet has the midline above center, and half has the midline below center.

ABCDEFGHIJKLMNOPQRSTUVWXYZ

Fig. 5-14 Lettering styles used by professional designers

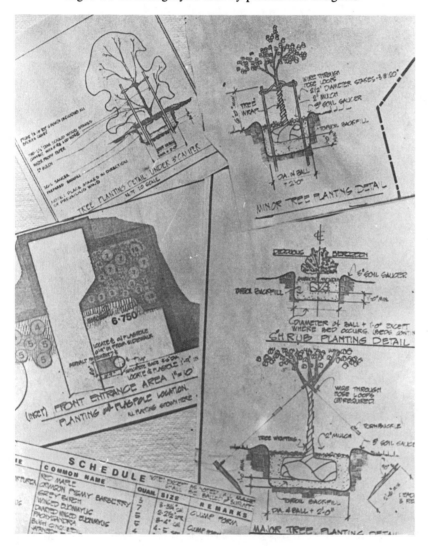

styles used by some professional landscape designers are shown in figure 5-14. Sense the life and vitality of these letters. There is not a dull, mechanical sameness about them as there is with letters produced by a lettering guide; neither is there a childish quality about them.

Once the four basic styles of free-hand lettering have been studied and mastered, students should look for examples of interesting lettering based upon these styles and try to duplicate them. One of the best ways to learn a new style is by copying those that you admire. Since a copy is never exact, a customized lettering style usually develops which then becomes the new designer's own.

Deliberate Line Width Variation

In the section describing the free-hand styles, it was mentioned that line width should be consistent. Such is the case when a round-lead pencil or technical pen is being used. However, careful study of some of the styles shown in figure 5-14 reveals noticeable variations in the line widths within a single letter. These letters are styled by using a *wedge* or *chisel point,* not a rounded lead. The wedge point has a beveled end, similar to a chisel, figure 5-15.

By using a wedge point, both broad and narrow lines can be made with the same pencil, figure 5-16. A simple version of this style can be developed by holding the wedge point in a fixed position: vertical strokes create a wide line and horizontal strokes create a thin line. Further practice with the chisel tip held at different angles gives the designer one more variation upon the basic block lettering style.

The following are the advantages of a good free-hand lettering style:

- There is no initial or recurring expense for supplies
- It can be used with all drawing media, such as pencils, technical pens, and felt-tip pens
- The letters have the human touch not found in press-on or stenciled letters. They look handcrafted, not manufactured. Thus, the finished design has a customized appearance.

The disadvantages of good free-hand lettering are almost non-existent. However, free-hand lettering does not look good on the same drawing where lettering guides have been used. The contrast between the two methods is not attractive.

SELECTING THE APPROPRIATE LETTERING METHOD

The time and expense involved in the lettering of a landscape plan are determined by the type and importance of the work being lettered. As noted previously, press-on letters are regularly used on concept drawings and presentation plans where the graphic quality is important to the acceptance of the designer's proposal. The same may be said for the use of lettering guides. If the potential value of the project warrants the time needed to use the lettering guides, then their use is justified.

Free-hand lettering is the best choice when the potential value of the project is such that the expense of press-on letters or the time it takes for stenciled letters cannot be justified. Working drawings vital to the installation of the landscape do not need

Fig. 5-15 Wedge-point drawing pencil

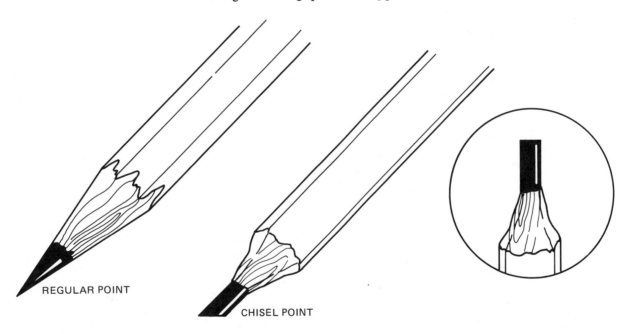

REGULAR POINT

CHISEL POINT

Fig. 5-16 Wedge-point lettering

THIS IS A WEDGE-POINT STYLE

stylized lettering, only complete information. They are almost always lettered freehand.

It should not be assumed that free-hand lettering is inappropriate on important presentation work. When done well, it is not only appropriate but preferred by many landscape designers over the other methods. Some designers never use lettering guides, and only use press-on letters to highlight such features as the client's name and address or the name of the designer or firm.

CREATING WORDS

Since the purpose of the landscape plan is to communicate the designer's ideas and intent to the client and the landscape installers, the letters must be formed into words. To create words correctly, the letters must be spaced properly. Proper spacing is basically a visual skill; that is, if it looks good, it is probably correct.

The following guidelines may help to insure attractive word formation:

- The letters within a word should not be crowded; neither should they be spaced so far apart that gaps are created, figure 5-17.

- The spaces between words should be greater than the spaces between letters within a word, figure 5-18.

- The same style of lettering should be used throughout the plan.

- Uppercase letters (capitals) and lowercase letters (noncapitals) should not be mixed within a word or label, figure 5-19.

Fig. 5-17 Incorrect Spacing

Fig. 5-18 Correct spacing

Fig. 5-19 Mixing uppercase and lowercase letters is improper.

PRACTICE EXERCISES

A.   On a sheet of drawing paper, lay out two horizontal lines one-eighth inch apart. Following the examples shown in this unit, letter a complete alphabet in the basic block style. Space the midline equidistant between the two horizontal lines and strike 90-degree vertical reference lines before beginning. Letter the alphabet several times. Repeat any letters you find difficult as often as necessary.

B.   Repeat exercise A, altering the placement of the midline and/or the angle of the vertical reference lines to create the full alphabet in (1) distorted block, (2) slanted block, and (3) slanted, distorted block styles.

C.   Select the lettering style that seems most comfortable to you. Using a rounded lead point, letter the Objectives of this unit using that style. Make the letters one-eighth inch tall. Be careful to keep to the same style, and to space the letters and words properly.

D.   Repeat exercise C using a wedge point on the pencil instead of a rounded point.

E.   Select one of the lettering styles illustrated in figure 5-14 and letter the alphabet in that style. Make the letters one-quarter inch tall.

ACHIEVEMENT REVIEW

A.   List four characteristics of the three most common lettering methods that make them suitable for use on a landscape plan.

B.   List the three methods of lettering commonly used by professional landscape designers.

C.   Indicate if the following advantages or disadvantages apply best to press-on letters (P), lettering guides (G), or free-hand letters (F).

   1.   Time consuming for some to use

   2.   Expensive

   3.   Have the human touch

   4.   Result in excess, unused letters

   5.   Helpful to designers who do not do well at free-hand lettering and is not expensive after the initial investment

   6.   Assures that the pencil or pen line will always be the same width

D.   Answer briefly each of the following questions.

   1.   When a stylized, professional lettering style is needed, which method is best?

2. Lettering guides that are to be used with technical pens should have what type of edge to prevent them from smearing?

3. Two factors of construction determine the style of free-hand lettering. One is the angle of the slant of the vertical reference lines. What is the other factor?

4. Should the spaces between words be greater or smaller than the spaces between letters within a word?

unit 6

SYMBOLIZING LANDSCAPE FEATURES

OBJECTIVES

After studying this unit, the student will be able to

- symbolize all major features of the landscape.
- interpret a landscape plan.

When landscape designers are ready to present their ideas to clients, it is important that the plans be both attractive and easy to read. If a plan is accurate in suggesting how the finished landscape will look, there is a greater chance that the client will accept the designer's ideas.

To obtain an accurate picture, landscape designers use their drawing instruments to create symbols. *Symbols* are drawings which represent overhead views of trees, shrubs, and other items that make up a landscape. If the symbols are neatly constructed and all are drawn to the same scale, the final plan can be very impressive, figure 6-1.

When symbolizing landscape features, it is important to keep the symbol as simple as possible. At the same time, it should be suggestive of the actual appearance of the landscape feature.

LANDSCAPE SYMBOLS

Needled Evergreens. When thinking of the appearance of a pine or spruce tree or a prickly juniper, it becomes easy to understand the symbol used for the needled evergreen, figure 6-2. The symbol suggests the spiny leaves and rigid growth habit of these plants which are green throughout the year.

Deciduous Shrubs. The word *deciduous* indicates a plant that drops its leaves in the autumn. *Shrub* usually refers to a multi-stemmed plant. It differs from a tree, which usually has only one main stem or trunk. The symbol for deciduous shrubs, figure

Fig. 6-1 A completed landscape plan

Environmental Design For:
SCHOHARIE BOCES CENTER
Schoharie, New York

6-3, suggests their larger leaves and loose habit of growth. Notice that there is an *X* or dot in the center of this and other symbols. This mark locates the exact center of the plant and indicates the

Fig. 6-2 Needled evergreen symbol

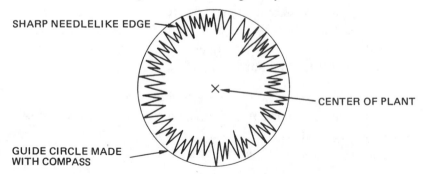

SHARP NEEDLELIKE EDGE

CENTER OF PLANT

GUIDE CIRCLE MADE
WITH COMPASS

Fig. 6-3 Deciduous shrub symbol

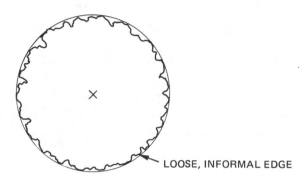

LOOSE, INFORMAL EDGE

point at which it is later to be set into the ground by the landscape contractor.

Deciduous Trees. Trees are usually larger than shrubs and play a different role in the landscape. The symbols for deciduous trees, figure 6-4, suggest their size and importance.

Broad-leaved Evergreens. Another group of plants keep their leaves all year (evergreen), but have larger, wider leaves than pines, hemlocks, or junipers. These are the broad-leaved evergreens. They are found in the warmer, milder regions of the country. Their mature appearance is less rigid than the needled evergreens and yet

Fig. 6-4 Deciduous tree symbols, one more complex than the other

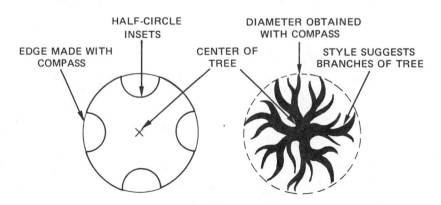

HALF-CIRCLE
INSETS

DIAMETER OBTAINED
WITH COMPASS

EDGE MADE WITH
COMPASS

CENTER OF
TREE

STYLE SUGGESTS
BRANCHES OF TREE

Fig. 6-5 Broad-leaved evergreen symbol

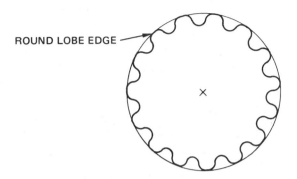

ROUND LOBE EDGE

somewhat more rigid than deciduous shrubs, as illustrated by the symbol in figure 6-5.

Vines. Trees and shrubs tend to grow in a circular manner. This is why the compass is so useful in forming their symbols. Vines, on the other hand, are wide and thin. The symbol for the vine is based more upon a sausage-shaped guideline, figure 6-6.

Ground Covers. The small trailing plants which cover the ground beneath shrubs and trees are called *ground covers.* There are many different types of plants which fall into this category. Some resemble vines; others are short and stalky. The symbol used to represent trailing ground covers does not show an individual plant, but rather a group of plants, figure 6-7.

Fig. 6-6 Vine symbol

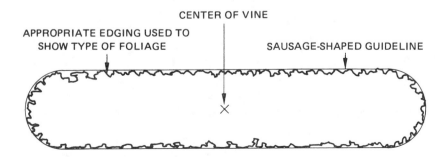

CENTER OF VINE

APPROPRIATE EDGING USED TO
SHOW TYPE OF FOLIAGE

SAUSAGE-SHAPED GUIDELINE

Fig. 6-7 Trailing ground cover symbol

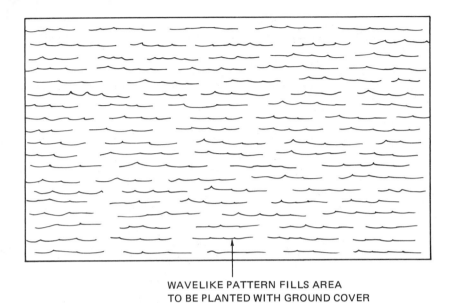

WAVELIKE PATTERN FILLS AREA
TO BE PLANTED WITH GROUND COVER

Construction Materials. The key to symbolizing constructed items of the landscape is to imagine how the item would appear if viewed from an airplane. That overhead view is how the drawn symbol should appear. Some common construction items are shown in figure 6-8. The student should practice drawing these and other symbols.

EXPLAINING AN IDEA WITH SYMBOLS

Once students have mastered the technique of symbol development, they can begin to use their skills to explain an idea. For example, figure 6-9 illustrates arrangements containing shrubs, trees, plants, and construction materials. Where plants touch (and the symbols merge) the plants are said to be *massed.* Where they do not touch, they are *side by side.* As illustrated in the drawings, it is sometimes necessary to indicate that one landscape feature is to be located beneath another. The symbol for the feature partially covered is drawn with a broken line where it passes beneath the more expansive feature. Hidden features are always shown as broken lines in symbolization.

Fig. 6-8 Some common construction symbols.

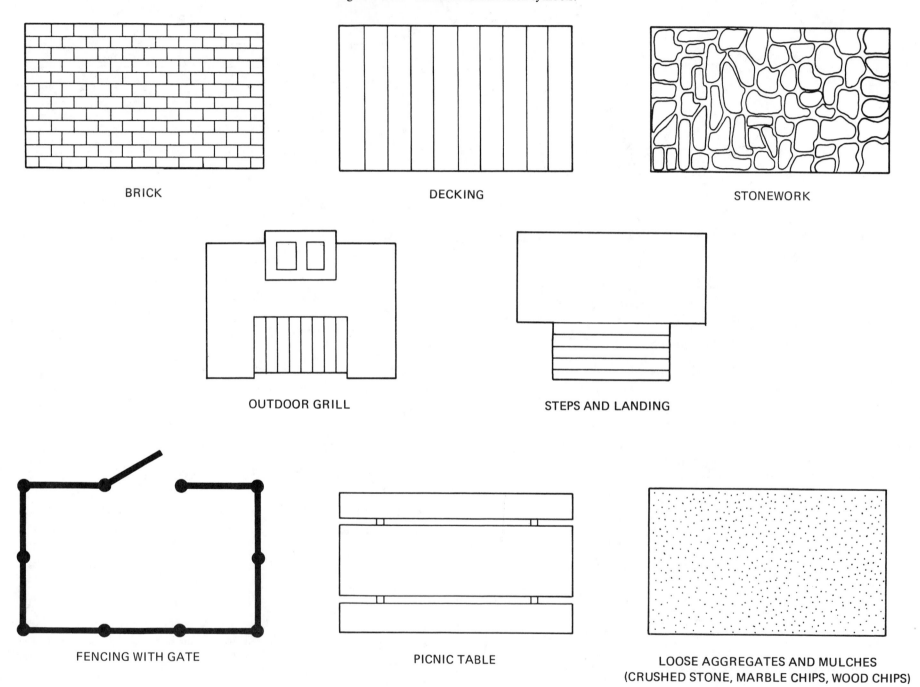

BRICK

DECKING

STONEWORK

OUTDOOR GRILL

STEPS AND LANDING

FENCING WITH GATE

PICNIC TABLE

LOOSE AGGREGATES AND MULCHES
(CRUSHED STONE, MARBLE CHIPS, WOOD CHIPS)

SYMBOLIZING TO SCALE

The final step in learning the technique of landscape symbolization is learning to draw the symbols so that they fit a certain scale. In developing a landscape plan, the landscape designer must draw all dimensions of buildings, property lines, and symbols to the same scale.

At this point, the student should demonstrate the techniques involved in drawing plant symbols to scale. The method is simple. Using the scale instrument, select the proper dimension for the compass (one-half the diameter desired). Form the guide circle with the compass and develop the appropriate symbol around the circle. The result will be a symbol which indicates the type of plant desired by the designer and the size it will be in the landscape.

Fig. 6-9 Symbols are grouped together to convey the designer's idea.

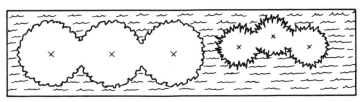

TWO SPECIES MASSED IN A BED WHICH
IS SURFACED WITH GROUND COVER.

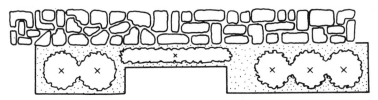

SHRUBS AND VINE AGAINST STONE WALL.
BED IS SURFACED WITH MULCH.

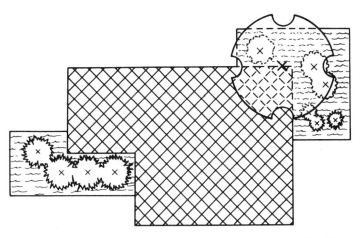

PATIO WITH PATTERNED SURFACE AND PLANTINGS.

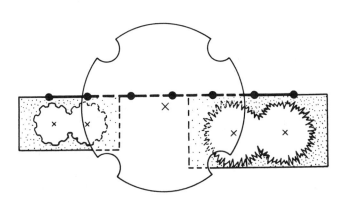

PLANTING BED AND FENCE PASSING BENEATH TREE.
BROKEN LINES ARE USED FOR HIDDEN FEATURES.

PRACTICE EXERCISES

A. Using the proper instruments, draw the symbols below to the given scale.

1. Deciduous tree, 25 feet wide, to the scale of 1″ = 10′.

2. Needled evergreen shrub, 10 feet wide, to the scale of 1″ = 10′.

3. Deciduous shrub, 8 feet wide, to the scale of 1″ = 5′.

4. Broad-leaved evergreen shrub, 6 feet wide, to the scale of 1″ = 5′.

B. Using drawing instruments, symbolize the features below to the proper scale.

1. Three massed needled evergreens (each 5 feet wide). Draw to the scale of 1″ = 5′.

2. A 25-feet wide deciduous tree with three deciduous shrubs each 8 feet wide beneath it. Draw to the scale of 1″ = 10′.

3. A patio, 10 feet x 15 feet, paved with brick. Draw to the scale of 1″ = 5′.

4. A side-by-side planting of the following, drawn to the scale of 1″ = 10′: three broad-leaved evergreens (each 5 feet wide), a needled evergreen tree (15 feet wide), four deciduous shrubs, massed, (each 6 feet wide), and a needled evergreen shrub (8 feet wide). Place the plants in front of a fence 65 feet long.

ACHIEVEMENT REVIEW

A. Study the completed landscape plan on page 51 and answer the following questions.

1. How many deciduous shrubs should be purchased for the plan?

2. How many deciduous trees are needed?

3. How many needled evergreens and broad-leaved evergreens are needed?

4. What is the difference in the symbolization used for ground covers and that used for mulches?

5. On which side of the house are the most trees located? Consult the plan's directional arrow to answer.

6. Knowing that the sun moves from east to west, which rooms of the house will be hottest in the afternoon?

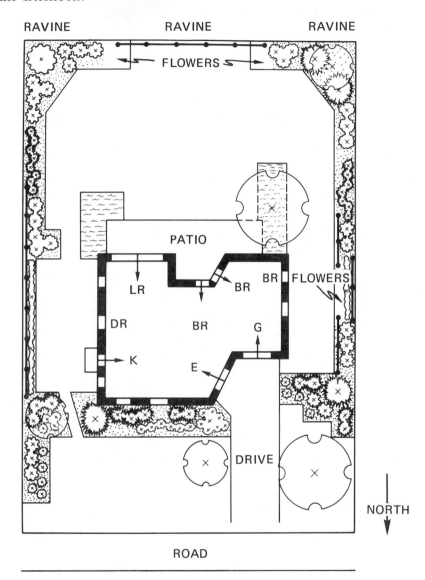

SCALE: 1″ = 20′

B. Select the best answer from the choices offered to answer each question.

1. What do landscape designers use most often to convey their ideas to their clients?

a. photographs   b. symbols   c. magazine pictures

2. The landscape plan is drawn to show what type of view?

    a. an overhead view   b. a ground-level view   c. an interior view

3. What term describes a plant which is not evergreen?

    a. dead   b. vine   c. deciduous

4. Why is an *X* or dot placed in the exact center of each symbol?

    a. to cover the hole made by the compass

    b. to locate the center of the plant and the point at which it will be installed in the ground

    c. to make the symbol more noticeable

5. To symbolize an evergreen shrub 8 feet in diameter, what should be the radius setting for the compass?

    a. 8 feet   b. 6 feet   c. 4 feet

SUGGESTED ACTIVITIES

1. Observe several different plants outside. Determine if they are needled or broad-leaved evergreens, deciduous trees or shrubs, vines, or ground covers. Make a tally to determine which types are found most commonly in your area.

2. Visit a nursery or garden center where a wide assortment of plants can be examined side by side.

3. Take drawing instruments outside. Select a planting near the school and draw it to the scale of 1″ = 10′. Use the correct symbol for each landscape feature.

4. Borrow a set of landscape plans from a landscape designer or landscape architect. Compare the symbols used with the ones illustrated in this unit. Note the difference between the symbols used in working plans and those used in presentation plans.

unit 7

EXAMPLES OF LANDSCAPE DESIGNS

The illustrations that follow are examples of landscape designs done by professionals currently working in the landscape industry. The designs include residential areas, small commercial sites, and recreational areas. These plans are the work of designers having college preparation ranging from two to five years. The designers are employed as landscape contractors, landscape architects, or recreational planners.

Students will gain several insights by studying the plans. First, each plan can be seen as a graphic explanation of the designer's ideas. Some are mainly concepts, with few precise details, aimed at selling the designer's proposals to a client. Others are detailed instructions of what, where, and how elements of the landscape are to be developed.

Second, notice how the graphics vary. Some of the plans are highly mechanical in appearance, due to the use of lettering guides and waxed press-on letters. Others are less rigid in appearance, due to free-hand lettering and a looser graphic style. All of the designs are of professional quality. The graphic technique used depends upon how much time the project warrants at a particular stage in its development, and how much competition there is among designers to gain a potential client.

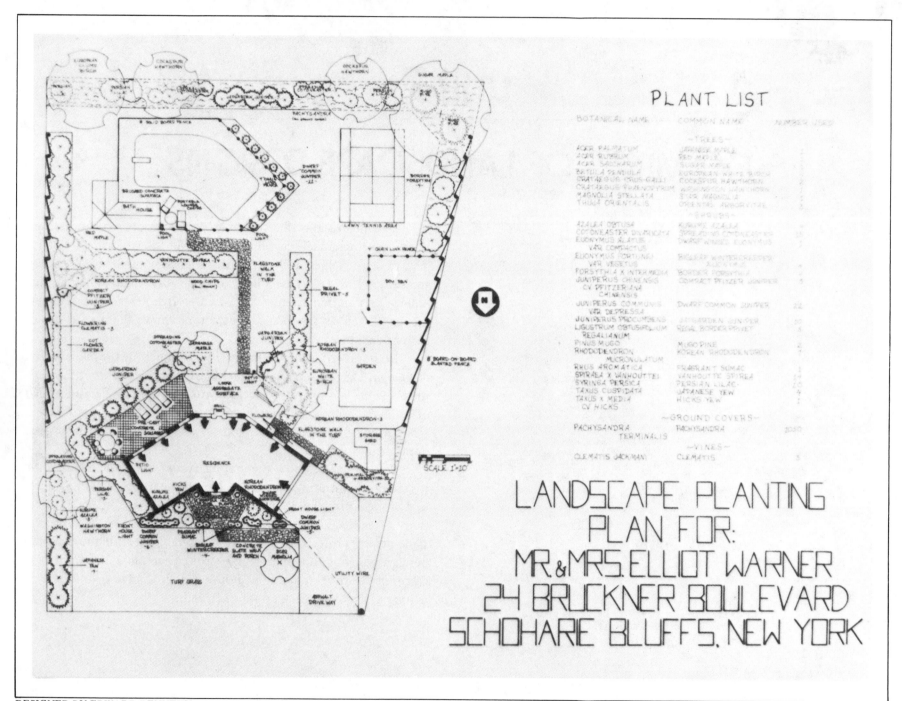

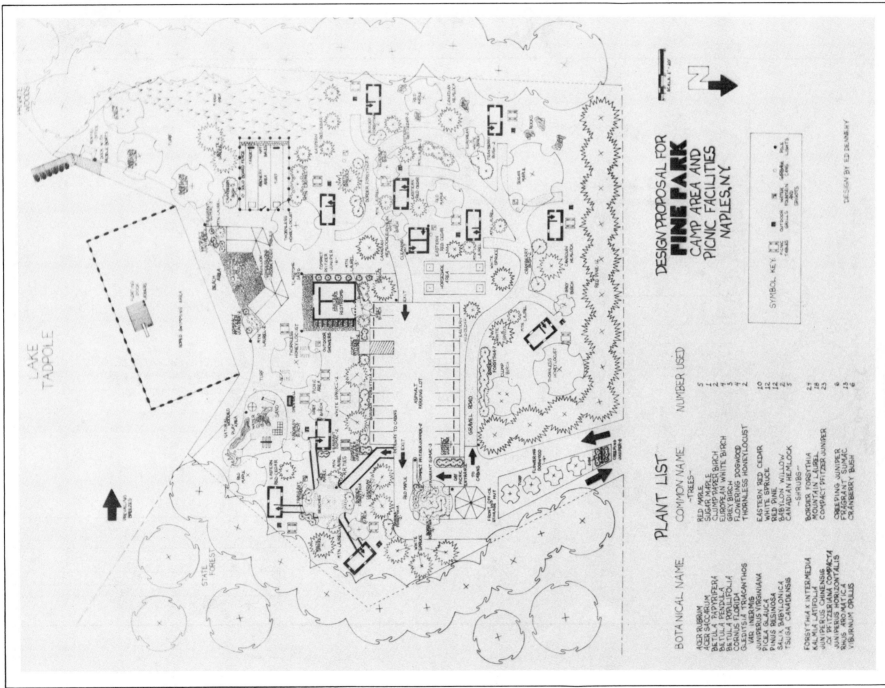

DESIGN PROPOSAL FOR
PINE PARK
CAMP AREA AND
PICNIC FACILITIES
NAPLES, N.Y.

DESIGN BY ED DENNEHY

SYMBOL KEY

PLANT LIST

BOTANICAL NAME	COMMON NAME	NUMBER USED
	—TREES—	
ACER RUBRUM	RED MAPLE	5
ACER SACCARUM	SUGAR MAPLE	1
BETULA PAPYRIFERA	CLUMP PAPER BIRCH	2
BETULA PENDULA	EUROPEAN WHITE BIRCH	1
BETULA POPULIFOLIA	GREY BIRCH	3
CORNUS FLORIDA	FLOWERING DOGWOOD	4
GLEDITSIA TRIACANTHOS VAR. INERMIS	THORNLESS HONEYLOCUST	2
JUNIPERUS VIRGINIANA	EASTERN RED CEDAR	10
PICEA GLAUCA	WHITE SPRUCE	12
PINUS RESINOSA	RED PINE	12
SALIX BABYLONICA	BABYLON WILLOW	1
TSUGA CANADENSIS	CANADIAN HEMLOCK	5
	—SHRUBS—	
FORSYTHIA X INTERMEDIA	BORDER FORSYTHIA	21
KALMIA LATIFOLIA	MOUNTAIN LAUREL	28
JUNIPERUS CHINENSIS CV. PFITZERIANA COMPACTA	COMPACT PFITZER JUNIPER	23
JUNIPERUS HORIZONTALLIS	CREEPING JUNIPER	8
RHUS AROMATICA	FRAGRANT SUMAC	15
VIBURNUM OPULUS	CRANBERRY BUSH	6

LAKE
TADPOLE

STATE
FOREST

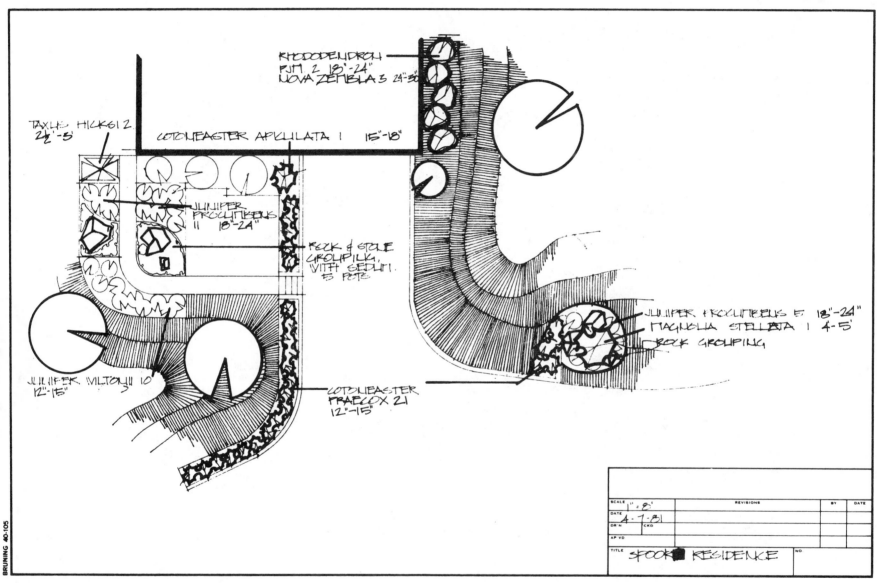

DESIGNED BY MICHAEL BOICE
ALBANY, NY

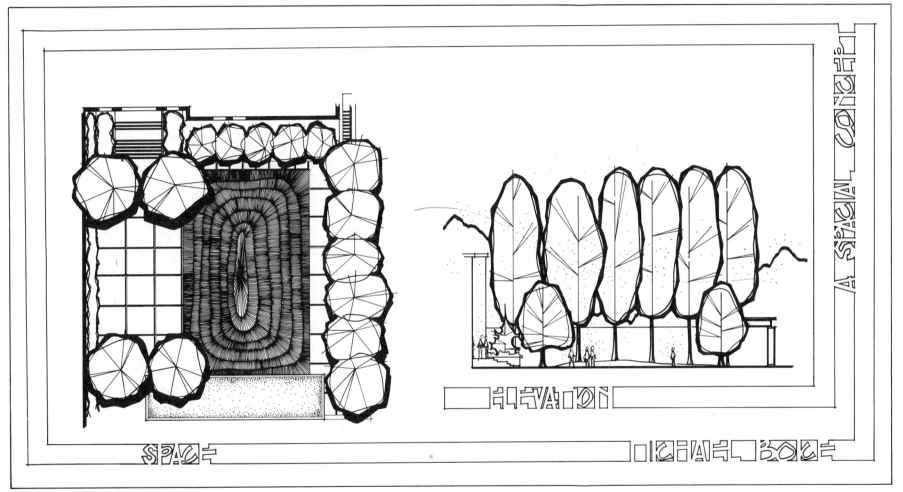

DESIGNED BY MICHAEL BOICE
ALBANY, NY

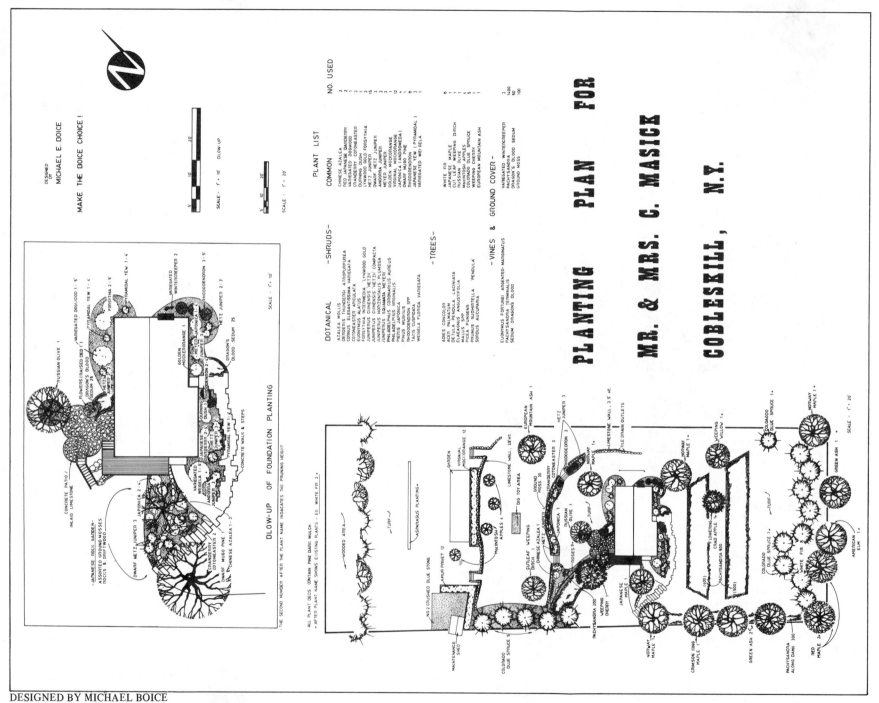

SITE PLAN

DESIGNED BY MARK MAGNONE
SCHENECTADY, NY

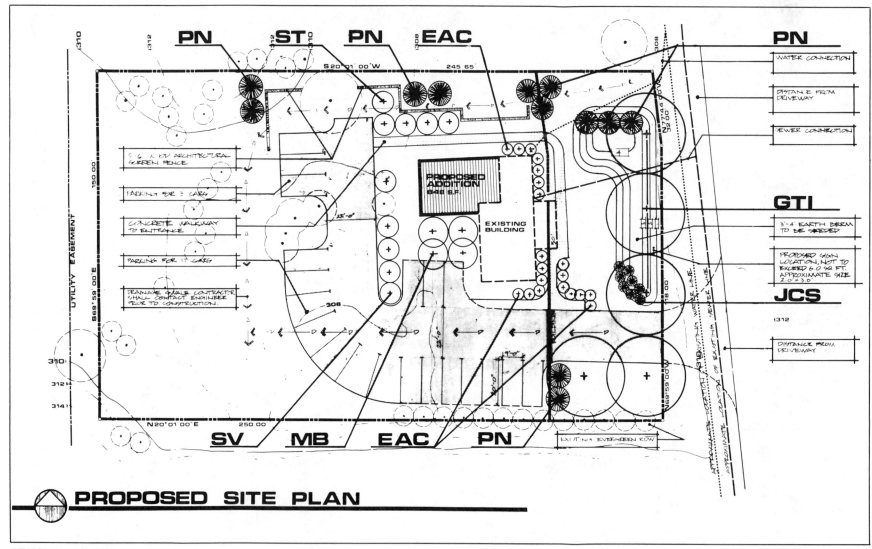

PROPOSED SITE PLAN

DESIGNED BY MARK MAGNONE
SCHENECTADY, NY

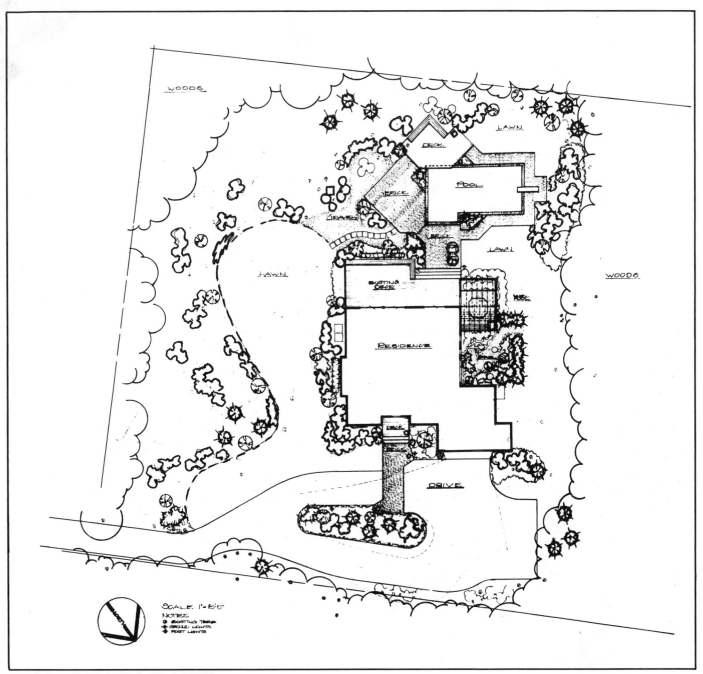

DESIGNED BY GOLDBERG AND RODLER
HUNTINGTON, NY

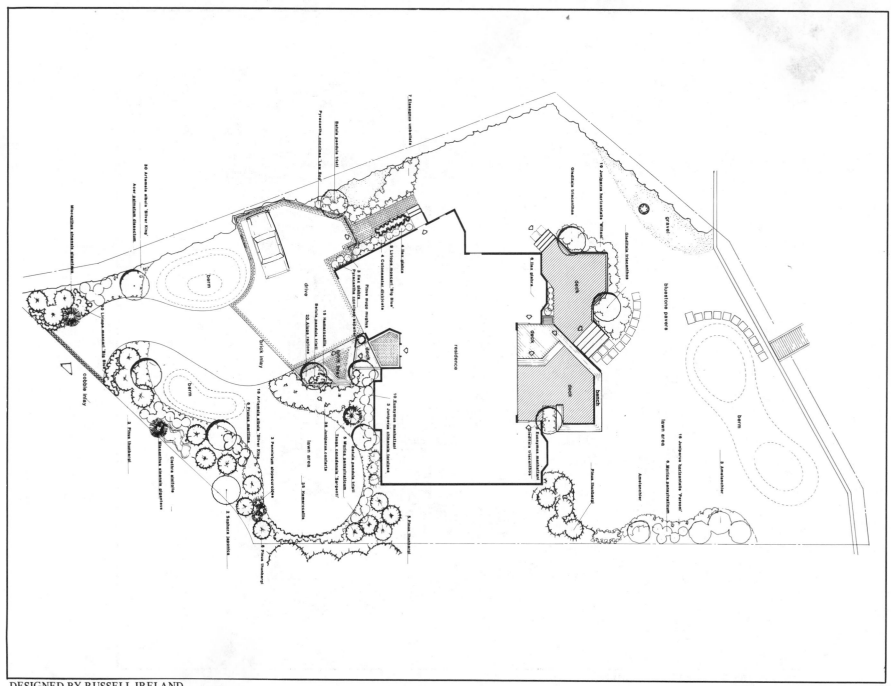

DESIGNED BY RUSSELL IRELAND
EAST NORWICH, NY

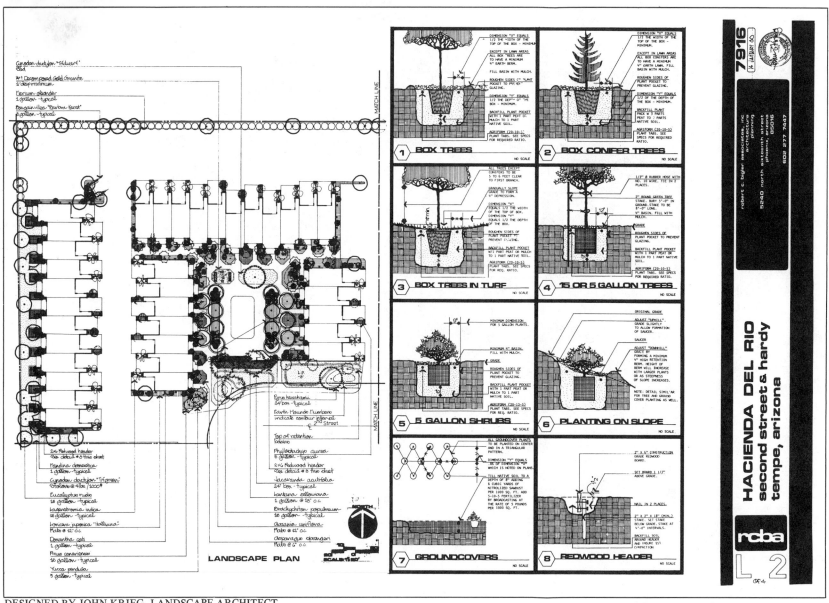

LANDSCAPE PLAN

DESIGNED BY JOHN KRIEG, LANDSCAPE ARCHITECT
ROBT. C. BIGLER ASSOC., ARCHITECTS
PHOENIX, ARIZONA

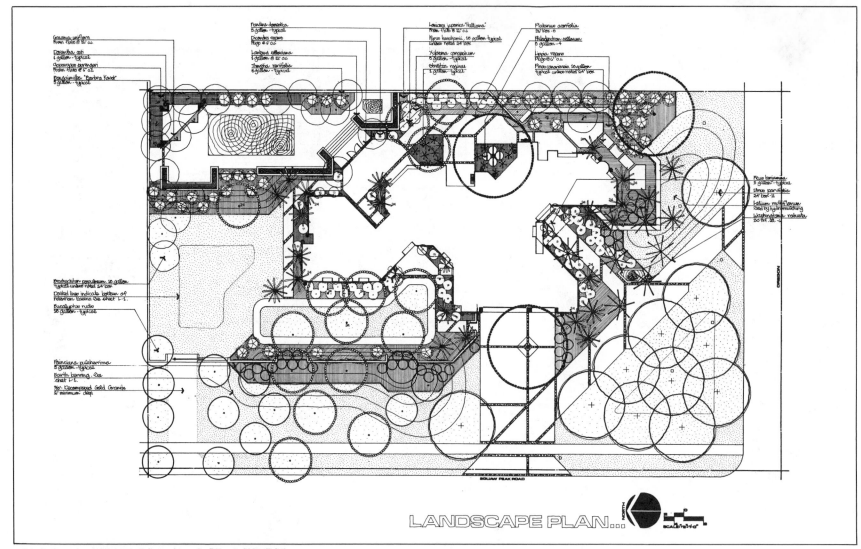

LANDSCAPE PLAN...

DESIGNED BY JOHN KRIEG, LANDSCAPE ARCHITECT
ROBT. C. BIGLER ASSOC., ARCHITECTS
PHOENIX, ARIZONA

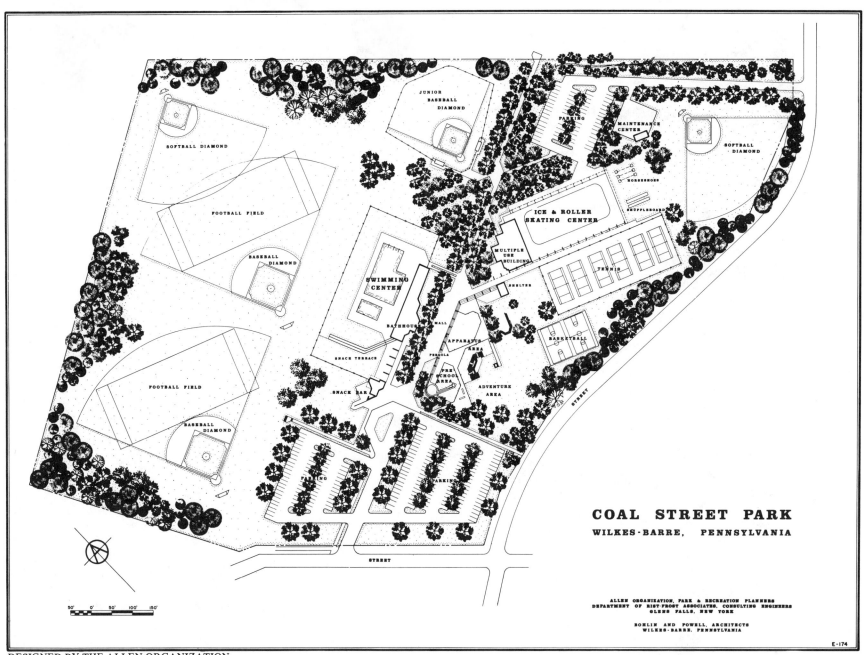

SOFTBALL DIAMOND

JUNIOR BASEBALL DIAMOND

PARKING

MAINTENANCE CENTER

SOFTBALL DIAMOND

FOOTBALL FIELD

HORSESHOES

SHUFFLEBOARD

ICE & ROLLER SKATING CENTER

BASEBALL DIAMOND

MULTIPLE USE BUILDING

TENNIS

SWIMMING CENTER

SHELTER

BATHHOUSE   MALL

BASKETBALL

SNACK TERRACE

APPARATUS AREA

PERGOLA

PRE SCHOOL AREA

FOOTBALL FIELD

SNACK BAR

ADVENTURE AREA

STREET

BASEBALL DIAMOND

PARKING

PARKING

STREET

50' 0' 50' 100' 150'

COAL STREET PARK

WILKES-BARRE,   PENNSYLVANIA

ALLEN ORGANIZATION, PARK & RECREATION PLANNERS
DEPARTMENT OF RIST-FROST ASSOCIATES, CONSULTING ENGINEERS
GLENS FALLS, NEW YORK

BOHLIN AND POWELL, ARCHITECTS
WILKES-BARRE, PENNSYLVANIA

E-174

DESIGNED BY THE ALLEN ORGANIZATION
GLENS FALLS, NY

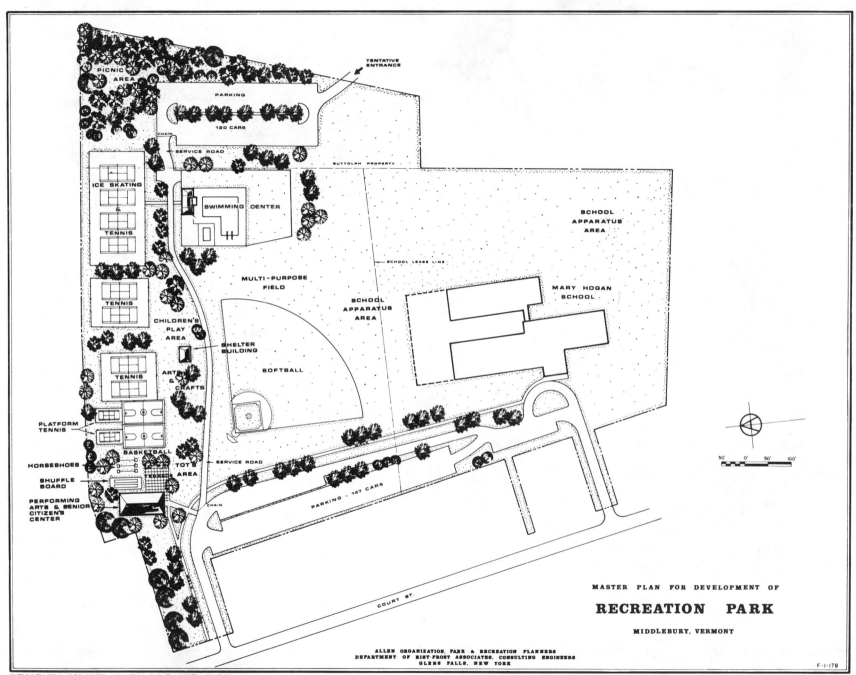

TENTATIVE ENTRANCE

PICNIC AREA

PARKING

120 CARS

CHAIN

SERVICE ROAD

SUTTOLPH PROPERTY

ICE SKATING & TENNIS

SWIMMING CENTER

SCHOOL APPARATUS AREA

SCHOOL LEASE LINE

TENNIS

MULTI-PURPOSE FIELD

SCHOOL APPARATUS AREA

MARY HOGAN SCHOOL

CHILDREN'S PLAY AREA

SHELTER BUILDING

ARTS & CRAFTS

SOFTBALL

TENNIS

PLATFORM TENNIS

BASKETBALL

HORSESHOES

SHUFFLE BOARD

TOT'S AREA

SERVICE ROAD

CHAIN

PERFORMING ARTS & SENIOR CITIZEN'S CENTER

PARKING - 147 CARS

COURT ST.

50' 0' 50' 100'

MASTER PLAN FOR DEVELOPMENT OF

RECREATION PARK

MIDDLEBURY, VERMONT

ALLEN ORGANIZATION, PARK & RECREATION PLANNERS
DEPARTMENT OF RIST-FROST ASSOCIATES, CONSULTING ENGINEERS
GLENS FALLS, NEW YORK

F-1-178

DESIGNED BY THE ALLEN ORGANIZATION
GLENS FALLS, NY

unit 8

ANALYZING THE LANDSCAPE SITE

OBJECTIVES

After studying this unit, the student will be able to

- define the word *site,* and explain its significance in the development of a landscape.
- list the features by which a site can be evaluated.
- describe the limitations that the terrain imposes upon human activities.
- understand the basic concepts of land grading.

THE SITE

The word *site* refers to a piece of land that has the potential for development. The evaluation of a site involves any or all of the following factors:

- A comparison of the characteristics of two or more possible land areas
- An analysis of the features of a single piece of land
- A comparison of the potential uses for a site
- Selection of a site, and a final detailed analysis of it

Site comparisons and analyses are part of the responsibilities of professional *site planners.*

The following examples illustrate how site analysis functions in the landscape process.

Comparing Sites

A religious organization plans to build a facility which will serve as a summer camp for its members and their families. The facility is intended to provide space for 30 residential cabins, a center for worship services, assorted woodland recreational areas, adequate parking, and all necessary public utilities. Privacy is also deemed to be a necessity.

Two sites are available, both of which are within the financial means of the organization. The principal characteristics of the two sites are as follows.

Site A

- The site consists of 500 acres; 400 are wooded and rolling, 100 are cleared and level.
- The site adjoins a major state highway.
- Land adjacent to the site is presently wooded, but is zoned commercial.
- A small stream runs through the property.
- Utility lines are already installed near the property.
- The soil is heavy clay, and one section of the wooded acreage has major rock outcroppings.

Site B

- The site consists of 550 acres, all of which are wooded and rolling.
- The site is near a state highway, but actually adjoins a secondary road.
- Land adjacent to the site is state forest land which is not likely to be rezoned.
- The property has no surface water, but it borders on a large lake.
- Utility lines are farther from the site than they are in Site A.
- The soil is loamy to sandy, and without rock formations.

Once the characteristics are noted, the site planner can begin a comparative analysis to determine which site will best satisfy the client's needs. This process is not like fitting a pair of shoes or solving a puzzle. There is never a perfect solution. The 100% perfect site does not exist among the available choices. Site analysis often results in reshaping the client's needs. Though failing to meet certain needs, the analysis may introduce new possibilities of the land that had not occurred to the client originally.

In this example, neither site is totally right nor wrong for the organization. Site A would permit the construction of cabins and other buildings, plus parking lots, on the 100 acres of cleared flat land, without the need for extensive and costly clearing and grading of the land. Site B would need to be cleared and leveled before building could begin. Site A is reached more easily than Site B.

However, the better site could depend upon the client's attitude toward remoteness and privacy.

Both sites offer sizeable woodlands, but the commercial zoning of the land around Site A could result in a great change in the character of the area over a period of years. Site B, surrounded by state-owned forest land, is unlikely to change as much as Site A, if at all, through the years. As a financial investment, Site A, because of the nearby commercial zoning, would increase in value faster than Site B. The preferable site could depend upon the client's concern or lack of concern about the land as an investment.

Finally, the lake in Site B would probably be considered as a more desirable feature for water recreation than the small stream in Site A. Whether the client would favor Site B over Site A would depend upon the importance of the lake as a positive site feature, as compared to the negative features of Site B and the positive features of Site A.

Analyzing One Site

A farmer wants to utilize 400 acres of marginally productive farmland in a way that will increase its profitability. The farmer seeks suggestions for potential use of the site as a recreational facility.

The principal features of the site are as follows.

- The site consists of 400 acres; 200 acres are cleared and level, and 200 acres are wooded but open beneath the canopy. The wooded acres were used as a grazing woodlot for cattle. The two 200-acre portions are separated by a stream where the trout fishing is good.

- The stream flows into a nearby public reservoir where recreational boating and swimming are allowed.

- Small wildlife abounds on the site.

- The soil is somewhat rocky and drains well.

- One corner of the woodlot is swampy and contains a growth of vegetation rare to the area.

- The site receives substantial snow accumulation during the winter.
- The land around the site is agricultural in character.
- The area is zoned for agriculture and small business.
- The site and adjacent areas were the scenes of many local skirmishes during the Civil War.
- The site is a rich source of American Indian artifacts, as well as prehistoric fossils.

In this example, the site analysis becomes an itemized account of the property's characteristics. Most of them cannot be rated as positive or negative, since the client has not yet decided how to put the property to use. The site's characteristics of themselves may suggest the best use. The terrain is suitable for all types of camping. Canoeing and boating are possibilities. Nature studies or hunting could take place in the swampy areas and habitat (living and growing) areas. The historical and archaeological features could be developed as tourist attractions.

In short, the site analysis can help to suggest uses for the land that the client had not thought of before.

THE CHARACTERISTICS OF A SITE

No two sites are identical. Although some may appear to be quite similar, each has a combination of qualities that sets it apart from other properties. A landscape planner should have an organized method to assess the hundreds of characteristics pertaining to each site.

Separating the factors into categories is a logical place to begin. Some of the site's characteristics are *natural* features, while others are *man-made*. Other features are *cultural* (associated with human society). Still others are basically *physical* features. Some other characteristics are most important as *visual* features. Some features are unmistakably *positive* factors, and others are definitely *negative* in impact. Many have a *neutral* quality until they are judged in the context of the proposed design.

A checklist is one way to compile the factors of a site analysis, figure 8-1. In this example, the information, when filled in,

Fig. 8-1 Site analysis checklist

Site Analysis for the Property of

Client's Name _Mr. & Mrs. John Doe_
Client's Address _1234 Main Street_
Tucson, Arizona
Taken By _JCS_ Date Taken _April 12, 1982_

Site Characteristics	Physical Importance	Visual Importance	Pos. +	Neg. −	Neutral ?
NATURAL FACTORS					
Existing vegetation	2 SHADE TREES IN RA. 1 FL. TREE OFF S W REAR CORNER	ALL IN GOOD HEALTH AND ATTRACTIVE	✓		
Stones, boulders, rock outcroppings	NONE	NONE			
Wind, breezes	WESTERLY / GUSTY			✓	
Surface water features	NONE	NONE			
Groundwater	TOO DEEP TO BE USABLE				✓
Soil conditions	CALICHE LAYER			✓	
Birds and small game	✓	✓	✓		
Large game	NONE				
Existing shade	FAIRLY GOOD		✓		
Turf plantings	NONE OF QUALITY				✓
Terrain features	LEVEL				✓
Direct sunlight	LOTS OF IT / ALL DAY				✓
Off-site views		MOUNTAINS IN DISTANCE	✓		
Others					
Hardiness zone 9					
Soil pH 8.4					
Soil texture SANDY DOWN TO THE CALICHE					
MAN-MADE FACTORS					
Architectural style of building(s)		SPANISH			✓
Presence of outbuildings	NONE				✓
Existing patios	10X15 CONCRETE	NOTHING SPECIAL			✓
Existing walks, paths, steps, ramps	DRIVEWAY IN / WALKS IN / CONCRETE		✓		
Swimming pool	YES	OVAL / 30' LONG	✓		
Fountains, reflecting pools	NONE	NONE			
Statuary	NONE	NONE			
Fences, walls	NONE	NONE			
Existing lighting	NONE	NONE			
Off-site features		HOMES ON ALL SIDES			✓
Others FIRE HYDRANT	FRONT CORNER OF LOT			✓	
CULTURAL FACTORS					
Power lines (aboveground)	YES	UNATTRACTIVE		✓	
Power lines (belowground)	NO	NO			
Telephone lines	YES			✓	
Water lines	INSTALLED				✓
Historical features	NONE	NONE			
Archaeological features		ATTRACTIVE SPANISH STYLING	✓		
Nearby roadways	IN FRONT OF HOUSE				✓
Neighbors	GOOD RELATIONSHIP / ELDERLY ON N SIDE FAMILIES + KIDS ON E+W		✓		
Off-site benefits		GOOD DISTANT VIEWS	✓		
Off-site nuisances	WIND IS TOO STRONG			✓	
Zoning regulations	YES / PROTECTIVE		✓		
Nearby public transportation	NO				✓

will be both complete and concise. The checklist can then be taken to the drawing table to provide essential input for the design process.

The analysis checklist is a flexible tool for the designer. Items can be simply checked (✓) to indicate their presence, or additional notations can be made to indicate their size, direction, quality, state of repair, importance, prominence, possible uses, and so on. The more notations there are, the more helpful is the data in producing a landscape plan that will fit the site.

READING THE TERRAIN

The rise and fall of the land describes its *terrain*. The record of an area's terrain is its *topography*. Topography is charted and recorded nationally by the United States Geological Survey, in the form of topographic maps, figure 8-2.

The maps are drawn to a scale of 1 inch = 2,000 feet. The broken lines on the map are *contour lines*, representing a vertical rise or fall of 10 feet over the horizontal distance measured from the map's scale. Each contour line connects all of the points of equal elevation on that map, and is labeled to indicate its elevation. The vertical distance between contour lines, *the contour interval,* is always stated on the map (10 feet in the U.S. Geological Survey topographic maps). Steep slopes are identified by closely spaced contour lines. Gradual slopes are denoted by more widely spaced contour lines. See figure 8-3.

Fig. 8-2 U.S. Geological Survey topographic map

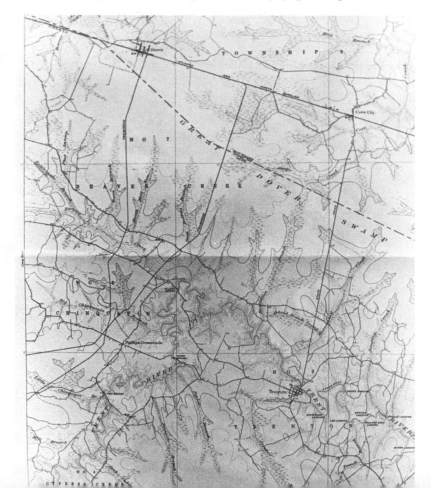

Fig. 8-3 The contour interval is 2 feet. Closely spaced contour lines represent a steep slope. Widely spaced lines show a gradual slope.

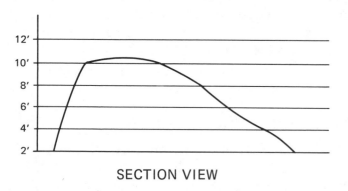

SECTION VIEW

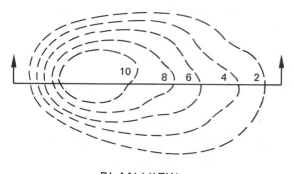

PLAN VIEW

For large sites, the U.S. Geological Survey maps may provide satisfactory data for a designer to use in planning the landscape. These maps are available for most areas of the United States, and may be purchased at a nominal cost from regional offices of the Survey. For smaller sites, the scale of the map may need to be 1 inch = 50 feet, or smaller, and the contour interval as precise as 1 foot between lines. To obtain such precise data, the landscape planner may have to hire a private surveyor.

Figure 8-4 illustrates some of the land forms recognizable from a topographic map. In order to interpret the map fully, students should know the following points regarding contours and contour lines:

- Existing contours are always shown as broken lines.

- Proposed contours are always shown as solid lines.

- Contours are labeled either on the high side of the contour or in the middle of the line.

- Spot elevations are used to mark important points.

- Contour lines neither split nor overlap (except in overhangs).

- Contour lines always close on themselves. The site map may not be large enough to show the closing, but it does occur on the land.

- Run-off water always flows downhill along a line that is perpendicular to the contour lines.

Once the contours of a site are known and plotted, then slopes can be measured and analyzed. *Slopes* are measurements that compare the horizontal length (measured from the map's scale) to vertical rise or fall (as determined by the contour lines and contour interval). Slopes may be stated as ratios or percents. As a *ratio,* the horizontal space required for each foot of vertical change in elevation is commonly expressed as 3:1, 4:1, and so forth, figure 8-5. As a *percent,* the vertical distance is divided by the horizontal distance, and the answer is expressed as 33%, 25%, and so forth.

Fig. 8-4 A topographic map labeled to show different land forms

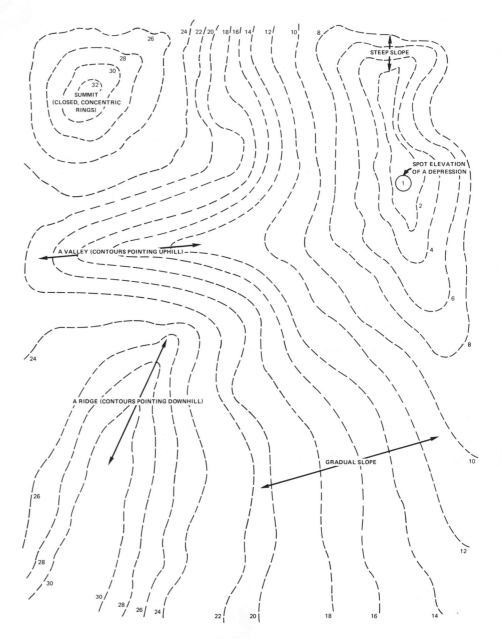

Fig. 8-5 Ratio of slope

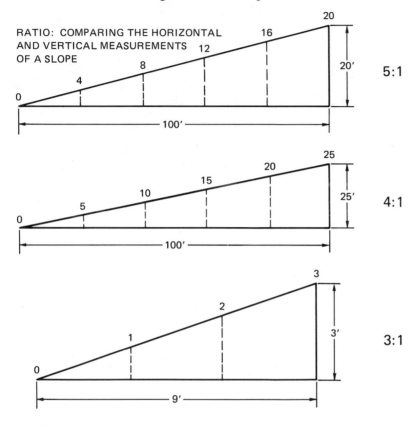

RATIO: COMPARING THE HORIZONTAL AND VERTICAL MEASUREMENTS OF A SLOPE

Fig. 8-6 Percent of slope

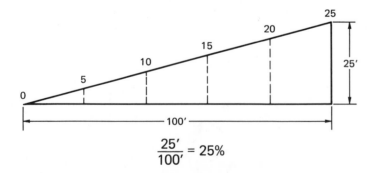

$$\text{PERCENT OF SLOPE} = \frac{VD}{HD} = \frac{\text{(VERTICAL DISTANCE)}}{\text{(HORIZONTAL DISTANCE)}}$$

$$\frac{25'}{100'} = 25\%$$

Another way to visualize percent of slope is to picture the slope extending along a horizontal distance of 100 feet. The vertical distance then becomes comparable to the percent of slope, figure 8-6.

THE NEED FOR TERRAIN INFORMATION

The ease or difficulty of development depends upon whether the land is level or rolling, rocky or sandy, forested or open. A study of the terrain also supplies answers to such basic questions as: Where does the surface water flow? Will water collect in puddles anywhere? What types of human activities can take place? Will grass grow on that slope? Can a car be parked safely on that slope?

Most human activities require that the land is flat or nearly flat. Land which is 5% or less in slope is perceived by users as flat. Flat land is the easiest terrain to develop, but it may be difficult to move off surface water. A slope of at least 1% is usually necessary to drain off surface water on turf and other planted landscape areas.

Human activities can usually take place on a slope of 5% to 10%, but users will sense the nonlevel footing. Land which slopes more than 10% may require alteration (grading) to make it more usable. Figure 8-7 lists the acceptable slopes for various landscape components.

GRADING THE LAND

When the terrain is not suitable for the activities planned for the site, it may be necessary to reshape it. The form of the land is changed by a process called *grading*. Grading can be as simple as one worker leveling and smoothing a small area of earth with a spade and a rake, and hauling away the leftover soil in a cart. It also can be so extensive that massive bulldozers and dump trucks are required to chew up and haul away entire mountains. Regardless of the extent of the project, grading is usually done for one of four reasons:

- To create level spaces for the construction of buildings
- To create the level spaces required for activities and facilities such as parking lots, swimming pools, and playing fields

Fig. 8-7 Recommended slopes for common landscape components

Recommended Slopes for Common Landscape Components

Landscape Component	Percent of Slope		Illustrated Example
	Allowable	Ideal	
Sitting areas, patios, terraces and decks	1/2% to 3%	1/2% to 2%	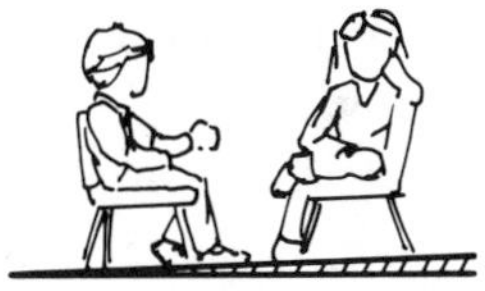 **2%**
Lawns	1% to 5%	2% to 3%	 **3%**
Walks	1/2% to 8%	1% to 4%	 **4%**
Driveways and ramps	1/2% to 11%	1% to 11%	 **10%**

Recommended Slopes for Common Landscape Components

Landscape Component	Percent of Slope		Illustrated Example
	Allowable	Ideal	
Banks planted with grass	Up to 33%	16% to 33%	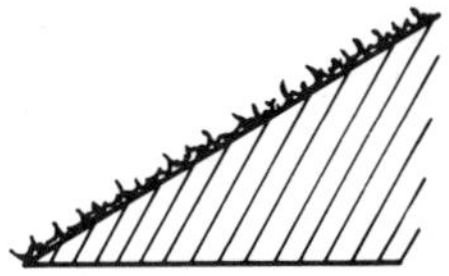 **30%**
Banks planted with groundcovers and shrubs	Up to 50%	20% to 33%	 **33%**
Steps	Up to 65%	33% to 50%	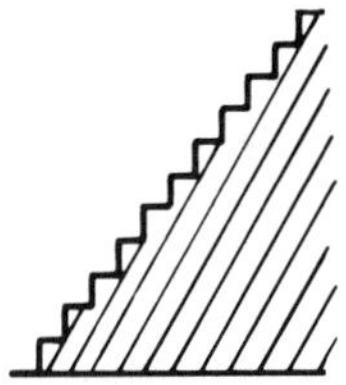 **50%**

- To introduce special effects into the landscape, such as better drainage, earth berms, tree wells, and ponds

- To improve the rate and pattern of circulation by means of better roads, ramps, tracks or paths

When earth is *removed* from a slope, the grading practice is called *cutting*. When earth is *added* to a slope, the practice is called *filling*. On a contour map, a cut is shown as (1) a solid line divert-ing from and then returning to an existing contour line, and (2) moving in the direction of a higher contour. A fill is shown as (1) a solid line diverting from and then returning to an existing contour line, but (2) moving in the direction of a lower contour line. See figure 8-8. A typical graded slope is illustrated in figure 8-9.

Since the grading process can involve the movement of tons of soil and rocks, designers should approach such specifications cautiously. In cut and fill operations, the soil that is removed

Fig. 8-8 Cut and fill as shown on a topographic map

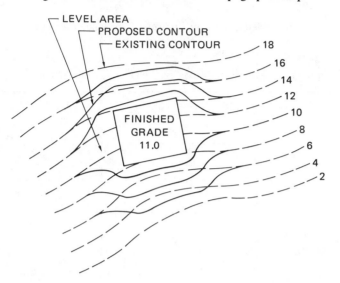

from the cut should be used to create the fill whenever possible. This practice minimizes the need for hauling. If possible, the top-soil layer should be stripped away and stockpiled before grading begins. The topsoil can then be spread over the finished grade before the site is replanted.

When land is graded, not only is the topsoil disturbed, but the surface water drainage and vegetation are disturbed as well. Water must drain away from buildings, not toward them. Freshly graded slopes must be stabilized to guard against erosion. The surface roots of valuable trees must be protected from the destruction of cutting and the suffocation of filling. Figure 8-10 shows a typical slope before and after grading. Figures 8-11 and 8-12 show common techniques for dealing with trees that exist prior to the grading of a site.

Fig. 8-9 Cross-sectional view of a typical graded slope

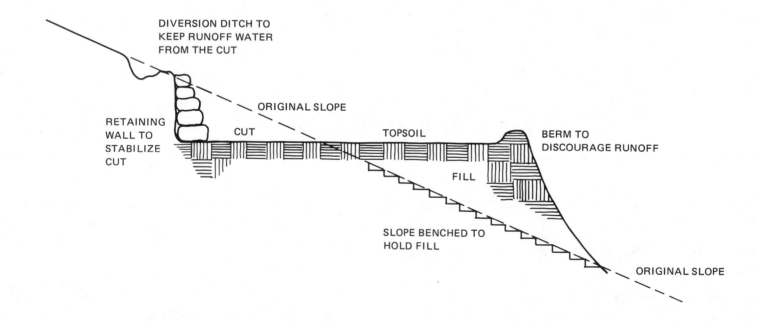

Fig. 8-10 A typical slope before and after grading

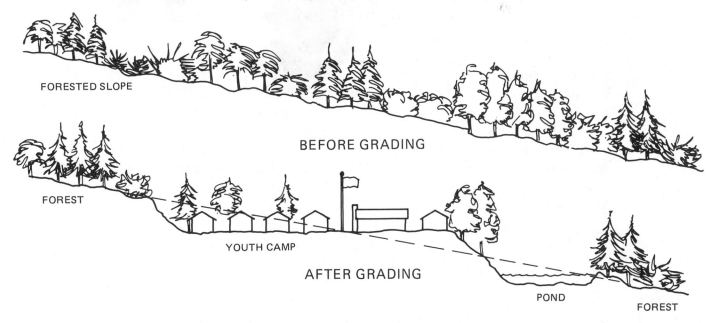

FORESTED SLOPE

BEFORE GRADING

FOREST

YOUTH CAMP

AFTER GRADING

POND

FOREST

Fig. 8-11 Raising the grade around an existing tree

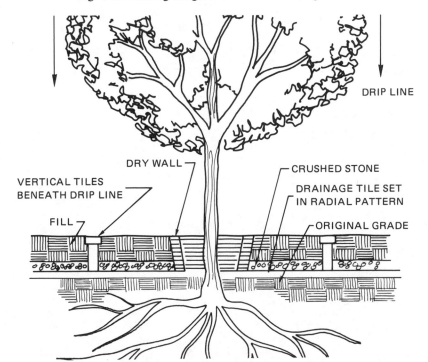

DRIP LINE

DRY WALL

CRUSHED STONE

VERTICAL TILES
BENEATH DRIP LINE

DRAINAGE TILE SET
IN RADIAL PATTERN

FILL

ORIGINAL GRADE

Fig. 8-12 Although the level of the lawn has been lowered, the tree's roots remain at the original level because of the retaining wall.

ACHIEVEMENT REVIEW

A.   Indicate if the site characteristics listed are natural (N), man-made (M), or cultural (C).

1. Condition of the turf
2. Nearness of public transportation
3. Rock outcroppings
4. Swimming pool
5. Terrain features
6. Off-site views
7. Buildings on the site
8. Style of the architecture
9. Existing shade
10. Prevailing breezes

11. Soil conditions
12. Historical features
13. Provisions for parking
14. Nearness of neighbors
15. Presence of wildlife
16. Traffic sounds
17. Zoning regulations
18. Presence of large, old trees
19. Existing lighting
20. Surface water patterns

B.   Define the following:

1. Terrain
2. Topography
3. Contour line

4. Contour interval
5. Slope

C.   Complete the following sentences that describe the characteristics of contours and contour lines.

1. Existing contours are always shown as _____ lines on topographic maps.

2. Proposed contours are always shown as _____ lines on topographic maps.

3. Contours are always labeled on the _____ side of the contour or in the _____ of the line.

4. Important points on topographic maps are marked by _____ .

5. A 3:1 or 4:1 comparison is the _____ of a slope.

D.   Label the parts of a typical cut and fill shown in the following diagram.

E.   Identify the land forms A through E on the following contour map.

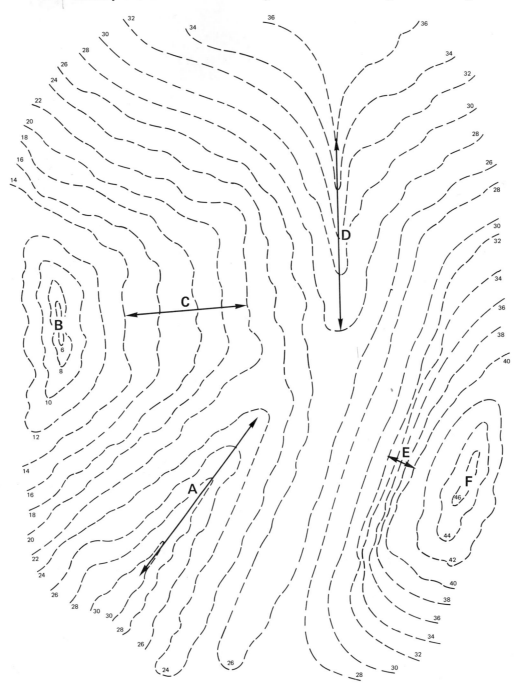

SUGGESTED ACTIVITIES

1. Fill several greenhouse flats with loamy soil and plant with grass seed. Tip the flats to create various slopes up to 50%. Water all flats with the same amount of water, applied from a sprinkling can. Record the erosion noted over the weeks at each degree of slope. Record also the quantity and quality of grass that grows.

2. Build a small contour model from information on a topographic map. Use cardboard of a thickness comparable to the contour interval on the map.

3. Using a sand table, the instructor can set up site situations that require alterations before landscaping can begin. Teams of students can alter the sites and discuss their reasoning with the class.

unit 9

THE OUTDOOR ROOM CONCEPT

OBJECTIVES

After studying this unit, the student will be able to

- identify indoor and outdoor use areas.
- list and define the features of the outdoor room.

Modern American homes may vary in size from three or four rooms to several dozen. Regardless of the simplicity or complexity of the home, the rooms are usually divided into four different categories of use, figure 9-1. The *public area* of the home is that portion which is seen by anyone coming to the house. The public area includes the entry foyer, reception room, and enclosed porch. The *family living area* includes those rooms of the house which are used for family activities and for entertaining friends. Rooms such as the living room, dining room, family room, and game room fall into this category. The *service area* includes those rooms of the house which are used to meet the family's operating needs. These include the laundry room, sewing room, kitchen, and utility room. The *private living area* of the home is used only by the members of the family for their personal activities. The bedrooms comprise the major rooms of this area. Dressing rooms, if present, are also included.

OUTDOOR USE AREAS

Just as a house is divided into different areas of use, so is the property around it. The public area of the landscape is the front yard. It is that portion of the landscape which is seen by everyone who drives or walks past the house. It is also the one area of the landscape through which everyone passes who enters the house. Ideally, the public area of the landscape connects with the public area of the home, figure 9-2. The public area should be large enough to place the house into an attractive setting, but not so large that a usable family living area is sacrificed.

Fig. 9-1 Examples of use area categories within the home.

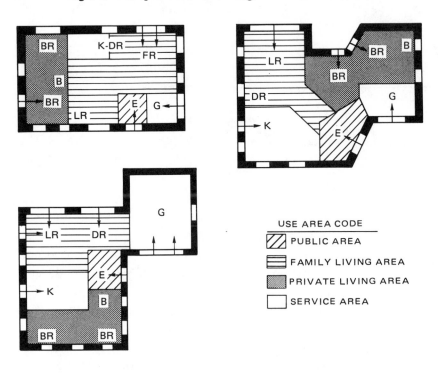

USE AREA CODE

◢◢ PUBLIC AREA

≡ FAMILY LIVING AREA

▒ PRIVATE LIVING AREA

☐ SERVICE AREA

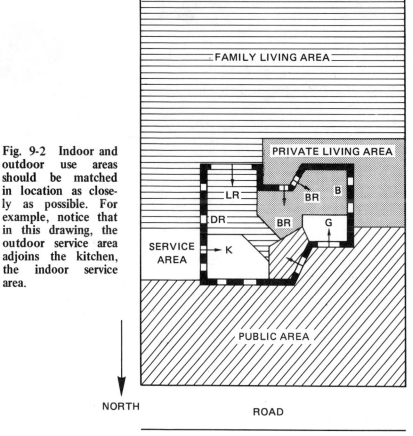

Fig. 9-2 Indoor and outdoor use areas should be matched in location as closely as possible. For example, notice that in this drawing, the outdoor service area adjoins the kitchen, the indoor service area.

The family living area of the landscape should also connect with the indoor family area whenever possible. This is often difficult because of the layout of some houses. The outdoor family living area is usually located toward the rear and often toward the side of the house. It is that portion of the landscape where the family relaxes and entertains guests. It includes space for patio, barbecuing, swimming, or whatever activities interest the family members. It is the largest of the outdoor use areas. Because of its location and its purpose, this portion of the landscape is seen by fewer people than the public area. It is developed for full or partial privacy. However, the quality of the design in both the family area and public area is important. They are both major areas of the landscape serving different purposes for the family.

The service area of the landscape plays a functional role for the family. It provides space to hang clothes, store garbage, house a dog, or grow a vegetable garden. Since it is used for service rather than beauty, the service area is usually screened from view. It is located near the kitchen or other indoor service room. The outdoor service area should be big enough to accomplish its purpose, but no larger.

All of the outdoor use areas described so far are most successful if they are connected directly to the indoor rooms they serve. A visual connection by way of a window is almost a necessity. A physical connection through a doorway is highly desirable. To be able to walk directly from the living room or family room into the outdoor family living area is a pleasure; to have to go through the kitchen or garage first is not so pleasant and may be a nuisance.

Fig. 9-3 Assigning landscape use areas is easier on some lots than on others.

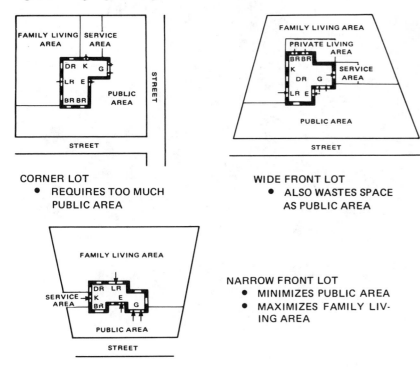

CORNER LOT
- REQUIRES TOO MUCH PUBLIC AREA

WIDE FRONT LOT
- ALSO WASTES SPACE AS PUBLIC AREA

NARROW FRONT LOT
- MINIMIZES PUBLIC AREA
- MAXIMIZES FAMILY LIV-ING AREA

Fig. 9-4 A typical outdoor room, containing wall, ceiling, and floor elements

The one outdoor use area which requires a direct physical link to the indoor room it serves is the private living area. This outdoor living area is developed for total privacy. The view of outsiders is screened. This is the area of the landscape where the family may sunbathe or relax in private. If the outdoor area is to serve the bedrooms of the house, it must be possible to reach the outdoor area without walking through another part of the house or yard. Therefore, a door off the bedroom is necessary if this area is being planned.

Figure 9-3 shows several different houses and lots divided into outdoor living areas. Note the sizes of the use areas and their relationship to indoor rooms, exits, and windows.

THE OUTDOOR ROOM CONCEPT

Once the use areas have been assigned to the property, the next step is to design and develop those areas into livable outdoor spaces. To accomplish this, it is helpful to think of the outdoor space as an outdoor room. In this way, students can apply some of what they know about indoor rooms to the outdoors.

The composition of the indoor room is the same whether it is simple or ornate, consisting of walls, a ceiling, and a floor. While the materials may vary from room to room, the basic structure remains the same.

The outdoor room also has walls, a ceiling, and a floor, figure 9-4. The materials are usually different from those used indoors, but they accomplish the same purpose. The *outdoor wall* defines the limits or size of the outdoor room. It can also slow or prevent movement in a certain direction. The walls of the outdoor room determine the vertical sides of the room in the same manner as the walls of an indoor room. Thus, outdoor wall materials should not be placed in the middle of the lawn, where a wall would not logically be located. Materials used to form outdoor walls may be natural (shrubs, small trees, ground covers, and flowers) or man-made (fencing and masonry).

The *outdoor floor* provides the surfacing for the outdoor room. The materials used for the outdoor floor might be natural surfacings such as grass, ground covers, sand, gravel, or water. They might also be man-made surfacings such as brick, concrete, patio blocks, or tile.

Fig. 9-5(A) Example of a public area

The *outdoor ceiling* defines the upper limits of the outdoor room. It may offer physical protection, such as an awning or aluminum covering, or merely provide shade, such as a tree. In temperate regions, a deciduous tree is an ideal ceiling material for placement near the home. It gives shade from the hot summer sun, and then drops its leaves in the winter to allow the sunlight through, which warms the house.

Figures 9-5 and 9-6 illustrate two use areas around a home landscape. Carefully analyze the development of these areas as outdoor rooms, noting the different materials used for the walls, ceilings, and floors. Also examine the corresponding landscape plan for each area.

Fig. 9-5(B) Plan view of the public area

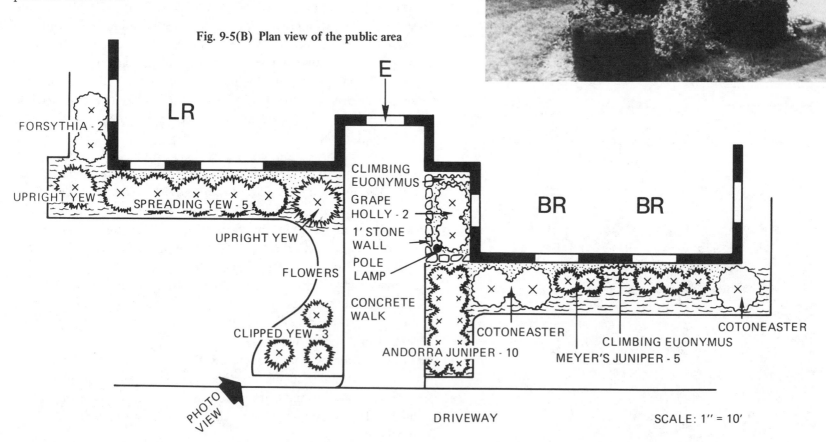

LR

E

FORSYTHIA - 2

UPRIGHT YEW
SPREADING YEW - 5

UPRIGHT YEW

FLOWERS

CLIPPED YEW - 3

CLIMBING EUONYMUS

GRAPE HOLLY - 2

1' STONE WALL

POLE LAMP

CONCRETE WALK

BR

BR

COTONEASTER

CLIMBING EUONYMUS

COTONEASTER

ANDORRA JUNIPER - 10

MEYER'S JUNIPER - 5

PHOTO VIEW

DRIVEWAY

SCALE: 1" = 10'

Fig. 9-6(A) Example of a family living area

Fig. 9-6(B) Plan view of the family living area

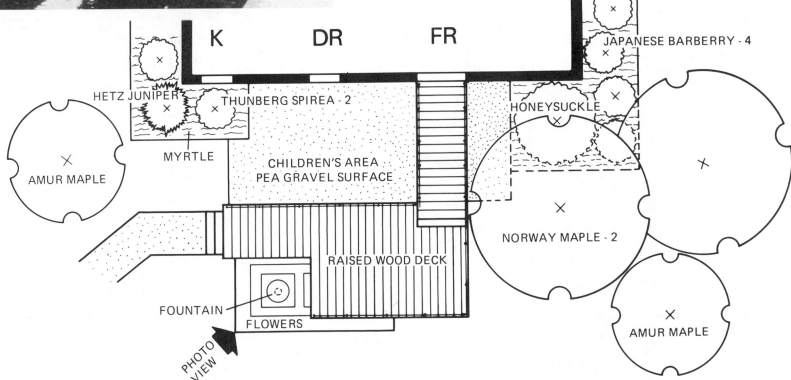

PRACTICE EXERCISES

A. Figure 9-7 shows three individual houses on their own lots. All windows, doors, and rooms are marked. Also provided is information on the family living in each house. Trace each house and its lot on tracing paper. Using a straightedge and pencil, divide each lot into the three or four use areas that seem appropriate for the family. With a scale instrument, determine the approximate square footage of each area you lay out. Is each area a practical size? Are the shapes such that rooms could later be developed if necessary? Is the family living area going to receive a good deal of sunshine? (It should be located on the south or west side of the property if possible.) Is the private living area receiving morning sun from the east? If it is intended as a morning use area, western afternoon sun is of little value.

Fig. 9-7

Family A:
a) An elderly couple whose children have moved away
b) Pet dog
c) Enjoy outdoor living, but are not very active
d) Garbage disposal

Family B:
a) A young couple with three children (ages 4, 6, and 8)
b) Pet dog
c) Enjoy outdoor living very much
d) Electric dryer

Family C:
a) Middle-aged couple with a 20-year-old daughter
b) No pets
c) Enjoy a moderate amount of outdoor living
d) Electric clothes dryer and garbage disposal

GREENWICH CIRCLE

LOT A

*The lot is level.
*There are neighbors on all sides of the property.
*Greenwich Circle is lightly traveled. It is used mainly as an access road for the properties adjoining it.

WOODS WOODS WOODS WOODS WOODS

110'

115'

SUNNY LANE

*The lot is level.
*The neighborhood is suburban.
*There is a house on the adjoining lot.
*Sunny Lane is a lightly traveled paved road.

LOT B

NORTH

SCALE: 1" = 40'

118'

94'

120'

100'

CENTRAL AVENUE

*The lot is level.
*There are neighbors on both sides of the lot.
*The rear of the lot connects with the backyard of the property on the next block.
*Central Avenue is a moderately busy street.

LOT C

B. Figures 9-8 and 9-9 illustrate two partially completed landscapes. Trace each landscape onto paper. Applying your knowledge of symbolization and the outdoor room concept, complete the landscape plans. As you insert symbols for trees, shrubs, and construction materials into the plans, be certain that each feature is playing the role of a wall, ceiling, or floor element.

Fig. 9-8 Partially completed landscape

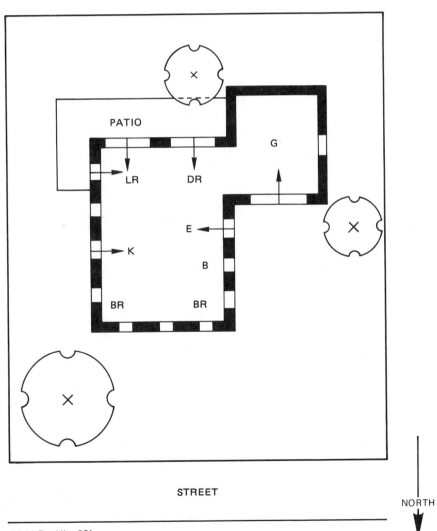

Fig. 9-9 Partially completed landscape

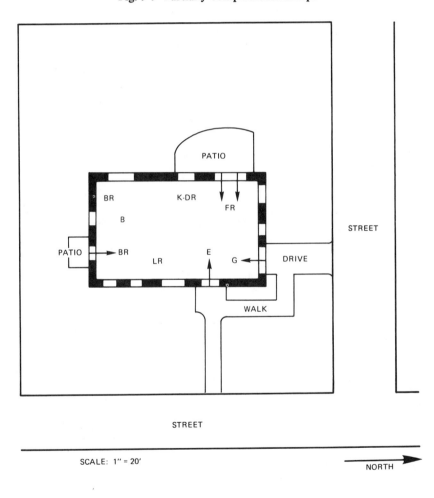

ACHIEVEMENT REVIEW

A.   Indicate in which use area the following activities occur.

Activity	Public Area	Family Living Area	Service Area	Private Living Area
picnicking				
welcoming guests				
hanging laundry				
breakfast, coffee				
badminton				
pitch and catch				
trash can storage				

B.   Indicate if the following materials are used for outdoor walls, ceilings, or floors.

Material	Outdoor Wall	Outdoor Ceiling	Outdoor Floor
brick wall			
shrubs			
crushed stone			
high-branching tree			
turf grass			
fencing			

SUGGESTED ACTIVITIES

1. Study various types of house floor plans. (These can be found in numerous magazines.) Which ones provide the best access to the outdoors through doorways? Which ones have only visual connection between the house and garden through windows?

2. Walk down a neighborhood street. Evaluate the public areas of the homes. Have the houses been placed in attractive settings? Are the wall elements where they should be or do they project into the center of the outdoor room?

3. Evaluate the property on which your school is located. Can different use areas be seen? Does the landscape appear to be a series of connecting outdoor rooms?

unit 10

DESIGNING PLANTINGS

OBJECTIVES

After studying this unit, the student will be able to

- identify the five principles of design and explain how they are applied to landscaping.
- arrange plant materials in linear and corner designs.
- explain the basic techniques involved in foundation planting design.

THE PRINCIPLES OF DESIGN

The landscape designer engages in a form of applied art, much as an architect or interior decorator does. Any form of art, including landscape designing, is guided by several basic principles. These *principles of design* have been applied by artists for centuries and are as relevant for today's students as they were for ancient Greeks.

Simplicity

The first principle of design is *simplicity*. To apply this principle, the landscaper must first realize that there may be complexities in the property before he or she even begins work. For example, a client may own a highly ornate home or desire the use of many different plants, flowers, and statues. In a case such as this, it is the task of the landscaper to develop simplicity in the property so that it provides a restful setting for outdoor living.

Simplicity may be incorporated in several ways. Repetition of the same species of plant, construction material, or color is one of the easiest ways to keep a landscape simple. It is not necessary, as some individuals believe, to use a great number of plants to have a well-landscaped home. Massing plants is another way to create simplicity. Massing can give the landscape a sense of unity

Fig. 10-1 Simplicity in plant arrangement. Grouping (B) allows the viewer's eye to move smoothly along the planting.

A. CHOPPY SILHOUETTE — NO MASSING

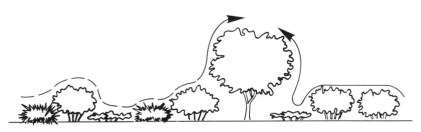

B. FLOWING SILHOUETTE — SPECIES MASSED

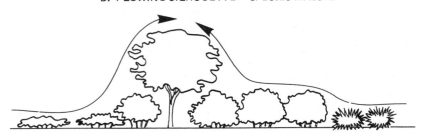

Fig. 10-2 Symmetric balance. One side of the outdoor room is a mirror image of the opposite side.

Fig. 10-3 Asymmetric balance. One side of the outdoor room has as much interest as the opposite side but does not duplicate it exactly.

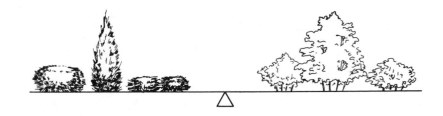

not otherwise possible, since the individual plants do not have to compete with one another for attention.

Simplicity in design does not necessarily suggest a boring or limited landscape. The design can be simple, yet still show full outdoor room development with a variety of plants and other landscape features. Figure 10-1 illustrates two *plantings* (groupings of plants), each containing the same plants. Notice how simplicity is achieved in planting (B) through the use of a smooth, flowing silhouette and the grouping of plant species.

Balance

The second principle of design is *balance.* To understand this principle, the student should imagine the outdoor room cut in half and placed upon a scale. If both sides of the outdoor room attract the eye of the viewer equally, the design is well balanced.

There are two types of balance useful in residential designing: symmetric and asymmetric. With *symmetric balance,* figure 10-2, one side of the outdoor room is planted and built exactly the same as the side opposite it. While symmetric balance is still used in landscape designing, it sometimes creates a formal appearance which may be undesirable. For this reason, asymmetric balance is more commonly used. As figure 10-3 shows, *asymmetric balance* creates the same amount of interest on both sides of the outdoor room, but does not create an exact duplication.

Proportion

Third among the principles of design is *proportion.* Proportion is concerned with the size relationship of the features of the landscape. In figure 10-4, a tree is shown between two houses. The tree is out of proportion with house (A), but is in proper proportion

Fig. 10-4 Proportion. Each element of the landscape must be in the proper size relationship with all other elements.

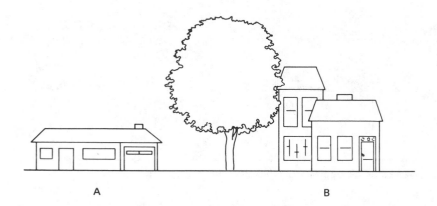

A   B

Fig. 10-5 Focalization. Plants are arranged in an asymmetrical step-down manner to move the viewer's eye toward the entry, the focal point.

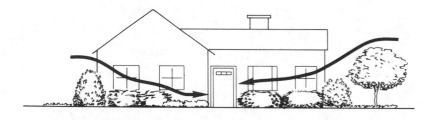

Fig. 10-6 Focalization draws the eye of the viewer to one major feature in each use area.

with house (B). Next to house (A), the tree looks gigantic, even threatening. With house (B), however, the tree seems appropriate and comfortable. The principle of proportion can be applied to plants, constructed features, and buildings. All must be in the proper size relationship with each other and, in a general sense, with the persons using the landscape.

Focalization

The principle of *focalization* is applied to every outdoor room in one or two selected spots. It is based upon the knowledge that when the human eye views a scene, it is attracted immediately to one feature, then gradually begins to take in the adjacent items. The feature which first attracts the eye is known as a *focal point*. It may draw attention through its shape, color, size, texture, sound, or motion. Examples of landscape focal points include *specimen plants* (highly attractive, unusual plants), flowers, statues, and fountains.

In the public area of the landscape, the most important feature is the front door of the house. Therefore, the focal point is already established and all other parts of the outdoor room are designed to bring the viewer's attention to the focal point, not compete against it, figure 10-5.

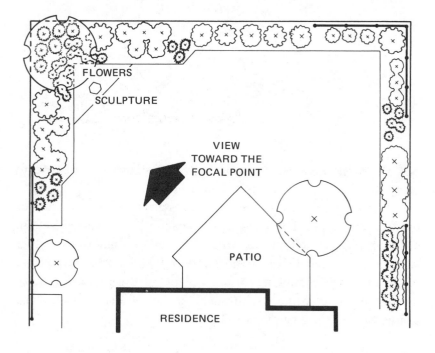

FLOWERS

SCULPTURE

VIEW
TOWARD THE
FOCAL POINT

PATIO

RESIDENCE

In other outdoor use areas, the designer may enjoy the freedom of creating the focal point. Figure 10-6 illustrates a backyard family living area with one corner developed as the major focal point. Many times, there is a temptation among new designers to include too many focal points in a design. It should be remembered that there is never more than one focal point per view.

Rhythm and Line

The final design principle is *rhythm and line.* This principle is used to create a sense of movement for the viewer's eye. Gently rolling bedlines and stepped plant arrangements are methods by which this sense of motion is created within an outdoor room. When walking from one use area into another, viewers should have the feeling that the design is transporting them from outdoor

Fig. 10-7 Rhythm and line. Two use areas are shown sharing a common planting whose bedline flows smoothly from one area to the next.

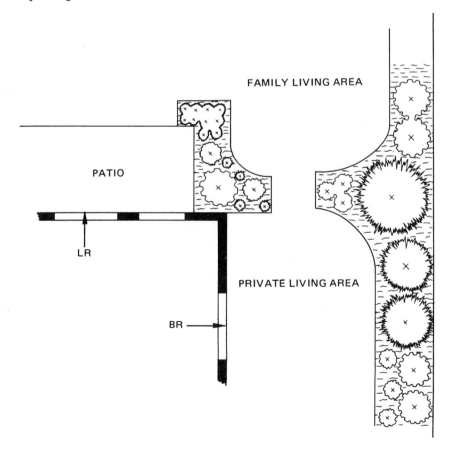

room to room. If the planting beds flow from one area to another, the principle of rhythm and line has been applied well, figure 10-7.

ARRANGING PLANT MATERIALS

No two properties should ever be regarded in exactly the same way, since different clients are being served in every case. Every landscape problem requires a fresh, new solution. Nevertheless, there are several basic ways of arranging trees and shrubs which can be applied to nearly all design problems.

A general rule of thumb for the student designer is that all shrubs belong in cultivated beds. The immediate reaction of a new designer to the above statement is often one of disagreement since in actuality, shrubs are not placed within beds in many instances. It is often wise, however, for students to avoid using what they see around them as examples when developing their landscape plans. The cultivated bed protects the shrubs from lawn mower damage by excluding the grass. It is mulched with peat moss, marble chips, wood chips, or similar material which helps to keep weeds out and moisture in. In this way, the cultivated bed helps to reduce the time and expense of landscape maintenance. In the examples of plant arrangements which follow, all shrubs and trees are placed in cultivated beds.

The Corner Planting

Defining the corners of the outdoor living room are the *corner plantings.* Depending upon how much privacy is desired, the corner planting may or may not be connected to other plantings which make up the walls of the outdoor room.

The corner planting bed has two parts, the *incurve* and the *outcurves,* figure 10-8. The incurve is the most desirable place for the location of a highly attractive plant (specimen plant) because it is a natural focal point. The plants in the outcurve should be selected and placed to draw attention even more strongly to the incurve.

The incurve plant is usually the tallest plant in the bed, figure 10-9. If the corner planting is not being used as the major

Fig. 10-8 Parts of the corner planting bed: the incurve and the outcurves.

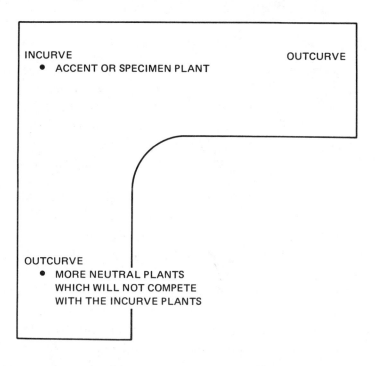

Fig. 10-9 In a corner planting, attention is drawn from the outcurves to the incurve by stair-stepping plants.

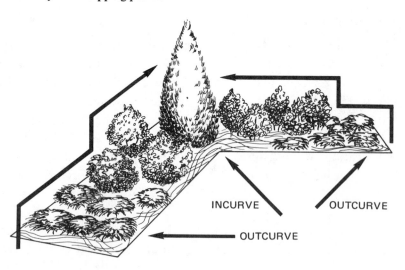

focal point of the outdoor room, a less attracting accent plant can be selected for the incurve. An *accent plant* has a greater attraction than the outcurve plants, but not as great an attraction as the focal point (in a different location).

There are many variations possible with the corner planting, figure 10-10. Shorter plants can be placed in front of taller plants, or a statue or bench might be used instead of a plant at the incurve. Whatever is done, it is important that the design remain simple. The number of plant species used should be limited to three or four unless the bed is exceptionally large.

Fig. 10-10 Three variations of the corner planting. Notice that in each example, the eye is drawn from the ends of the planting to the center and from the front to the rear.

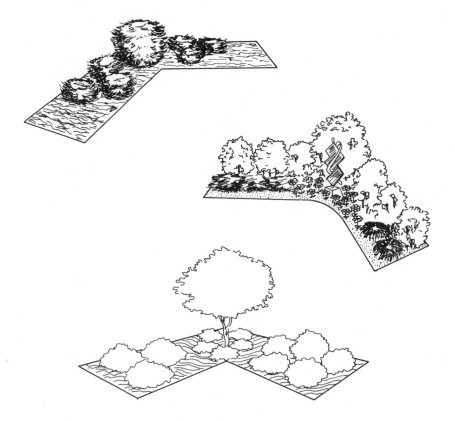

The Line Planting

The *line* or *linear planting* is the basic means of forming outdoor walls with plants. Depending upon the types of plants selected and their placement, the outdoor wall can accomplish many purposes. It can provide total privacy, partial privacy, or no privacy. The linear planting can also provide protection from the effects of the weather. It can create a view, block a view, or frame a view. Figure 10-11 illustrates some of the many functions of the line planting.

Fig. 10-11 Some functions of the line planting

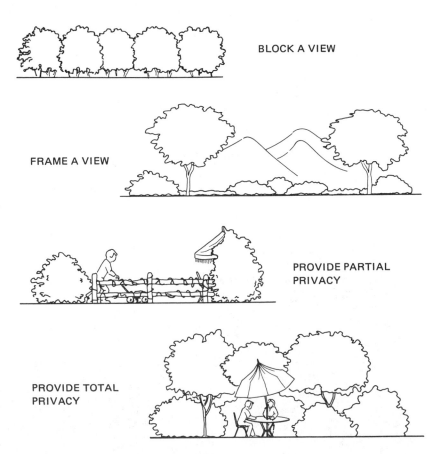

BLOCK A VIEW

FRAME A VIEW

PROVIDE PARTIAL PRIVACY

PROVIDE TOTAL PRIVACY

Fig. 10-12 An effective line planting consists of (1) only a few species, (2) massed groupings, and (3) staggered placement.

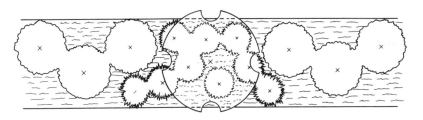

Fig. 10-13 The placement of low shrubs in front of taller ones adds a stepped effect to the line planting.

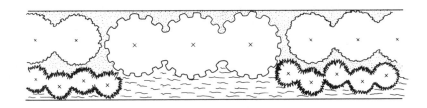

The line planting, like the corner planting, should be within a cultivated bed. The height and thickness of the outdoor wall depend upon the size and number of plants used. When arranging plants in this or any arrangement, remember to space the plants far enough apart so that they have sufficient room to grow to maturity. Crowded spacing creates unnecessary maintenance requirements.

Figure 10-12 illustrates the design approach to line planting. It is important that the plant species be limited and the plants grouped. The staggered placement of the plants makes the planting more interesting to the viewer's eye. Likewise, the small tree alters the silhouette of the planting, adding interest.

The line planting can be made more attractive by placing small shrubs in front of taller ones, figure 10-13. This creates a stepped effect, which can be further enhanced with ground cover placed below the smaller shrubs.

Skillful designing of the line planting requires practice and experience. Designers should try to avoid the monotony which develops when too few plant species are used, figure 10-14. Likewise, they must resist the temptation to use too many species, which destroys the simplicity of the design, figure 10-15.

The Foundation Planting

Many years ago, houses were built on high, unattractive foundations. The foundation planting developed as a way to hide the concrete block base of these older homes. Now, it is common practice to build homes so that the foundation does not show. The foundation planting is still used, though with a slightly different purpose.

Fig. 10-14 A monotonous view results when there is not enough variation in plant height or texture.

Fig. 10-15 The simplicity of a line planting is destroyed by too many plant species and too much variation in height.

Fig. 10-16 A common, unimaginative foundation planting. Upright shrubs at the corners of the house and spreaders beneath the windows create a rigid appearance.

Fig. 10-17 This foundation planting, with informal plant shapes, creates a modern appearance. The planting extends outward from the house to tie the structure more closely to the landscape.

Fig. 10-18 This planting, consisting of shrubs, ground covers and raised planters, carries the viewer's eye easily to the entrance of the home.

The modern foundation planting plays an important role in tying the house in with the rest of the landscape. No longer ending at the corners of the house, the foundation planting reaches out from the house toward the garden. It attracts the eye of the viewer and leads it toward the entry in the public area.

The beginning designer should avoid one common, unimaginative practice in the design of foundation plantings. This is the placement of spreading evergreens under the picture windows and upright pyramidal evergreens on the corners of the house, figure 10-16. By using greater imagination, the designer can create more attractive foundation plantings. Figure 10-17 shows the same house as figure 10-16, but with a more modern planting. The taller plants are placed at the corners and the height of the other plants gradually descends toward the entry. Figure 10-18 illustrates how a well-designed foundation planting can contribute to the beauty of a home.

PRACTICE EXERCISES

A.   Using the proper instruments and a drawing board, duplicate the corner planting bed shown below, drawn to the scale of 1″ = 10′. Design a planting for a viewer looking in the direction of the incurve. Select plants and dimensions from Chart A. Draw to scale and label all species used.

Fig. 10-19 Corner planting bed.

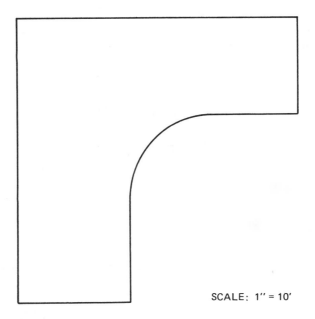

SCALE: 1″ = 10′

Chart A

Species		Width	Height
Redbud tree	(D)	20 ft.	25 ft.
Viburnum	(D)	10 ft.	12 ft.
Forsythia	(D)	10 ft.	9 ft.
Cotoneaster	(D)	5 ft.	5 ft.
Spirea	(D)	5 ft.	4 ft.
Grape holly	(BLE)	3 ft.	3 ft.
Andorra juniper	(NE)	3 ft.	1 ½ ft.
Myrtle	(G)	Vining	1 ft.

D: Deciduous   BLE: Broad-Leaved Evergreen
G: Ground Cover   NE: Needled Evergreen

B. With the proper equipment, duplicate the planting bed shown below. Assume that the viewer is located south of the bed and that there is an attractive mountain scene north of the bed. Design a planting which frames but does not block the view. Select plants from Chart A. Draw to the scale of 1″ = 10′.

Fig. 10-20 Planting bed

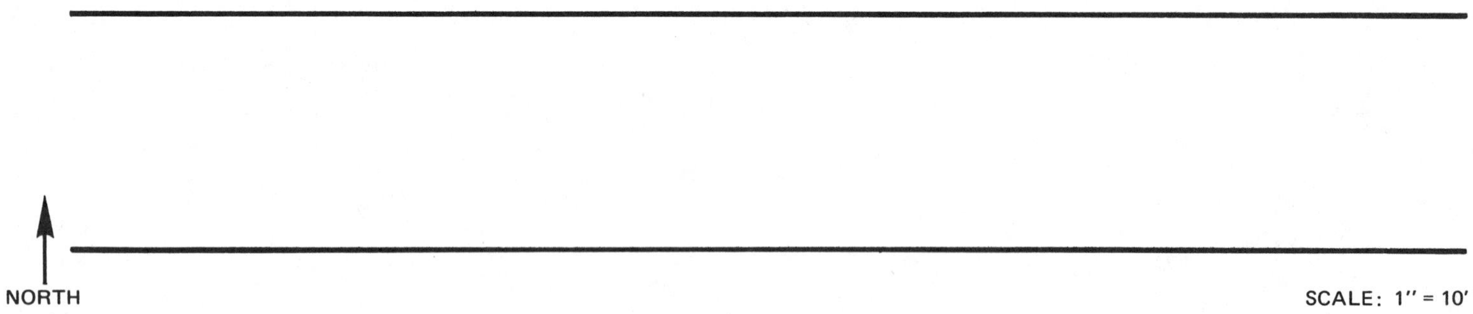

NORTH

SCALE: 1″ = 10′

C. Figure 10-21 illustrates the entry portion of a house. The garage and driveway are also shown. With the proper instruments, design the public area portion of the foundation planting. Use the plants in Chart A. Design to the scale of 1″ = 10′.

Fig. 10-21 Entry portion of house

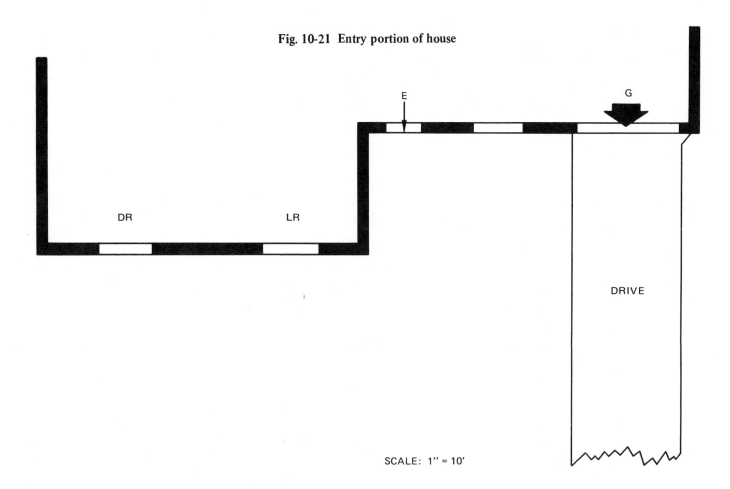

SCALE: 1″ = 10′

ACHIEVEMENT REVIEW

A. Fill in the blanks with the correct principle of design.

1. Drawing attention to the front door of a house is demonstrating the principle of _____ .

2. Repeating materials or plants in different areas of the landscape is demonstrating the principle of _____ .

3. The principle of design being ignored when a tree that will eventually grow to be 100 feet tall is planted next to a 15-foot house is _____.

4. A landscape which suggests smoothness and unity is demonstrating the principle of _____.

5. If one side of a landscape attracts the same attention as the opposite side, the designer has used the principle of _____.

B. Select the best answer from the choices offered to answer each question.

1. Which of the following is not a function of a good mulch?

   a. inhibits weeds   b. attracts insects   c. retains moisture

2. Which of the following plants would not work well at the incurve of a corner bed?

   a. vine   b. specimen plant   c. accent plant

3. Which is the proper way to space shrubs in a planting bed?

   a. so that the landscape appears completed the day it is installed
   b. so that shrubs are able to grow to maturity without crowding
   c. so that shrubs crowd together and require a great deal of pruning

4. Which is the function of the modern foundation planting?

   a. to hide the unsightly foundation of the modern house
   b. to prevent the house from being seen from the street
   c. to tie the house in with the rest of the landscape

SUGGESTED ACTIVITIES

1. Collect dried weed flowers or similar natural materials that resemble small trees and shrubs. Spray paint them green and arrange them on a styrofoam base cut into the shape of a corner or linear planting bed. See how many different ways the same bed shape can be designed.

2. Evaluate your home landscape plantings. How well have the plants been selected and arranged?

3. Evaluate the plantings of ten different home landscapes in a single neighborhood. Rate them on a 1 to 10 rating basis, with 10 representing an ideal landscape and 1 representing almost no development. Let 5 represent a very ordinary style, such as the one shown in figure 10-16. Average the scores to determine the rating for the neighborhood. Compare your average with other members of the class. If the neighborhood ratings vary greatly, determine the reasons why.

 # unit 11

COMPLETING THE LANDSCAPE PLAN

OBJECTIVES

After studying this unit, the student will be able to

- label a landscape plan.
- prepare a final landscape plan.

Since the plan is the major contribution of the designer to the landscaping process, it is important that it look professional. Generally, plans must be of better quality than clients could produce themselves. For this reason, the student designer should learn to incorporate some small but important final touches to complete the plan.

The finished design plan should be labeled throughout. This enables clients to read and understand the plan. It also aids the landscape contractor who is responsible for installing the design once it has been accepted by the client. There are several types of labels which appear on the finished plan. Included are symbol labels, directional arrow, scale, designer's name, client's name and address, and plant list.

Symbol labeling identifies each plant, construction item, installation detail, and building which appears on the plan. Nothing must be left unexplained. To avoid confusion, it is always best to label directly on or near the symbol. It may not always be possible to fit the entire label directly on the symbol; therefore, all of the exampes shown in figure 11-1 are acceptable. These methods of symbol labeling are most desirable because they allow for immediate reading of the plan.

Coding, figure 11-2, is sometimes necessary when the scale of the plan is $1'' = 20'$ or smaller. Coding is a less satisfactory method since it requires a separate listing to explain the code. Should the list be lost, the design is unreadable.

Following the name of a plant or its coded number is the number of plants to which that label applies, figure 11-3. By using

Fig. 11-1 Three acceptable methods of direct symbol labeling.

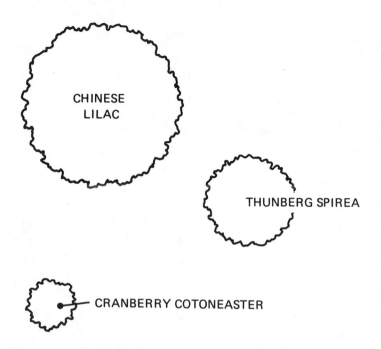

Fig. 11-3 When the label or code applies to more than one plant of that species, the number should be indicated.

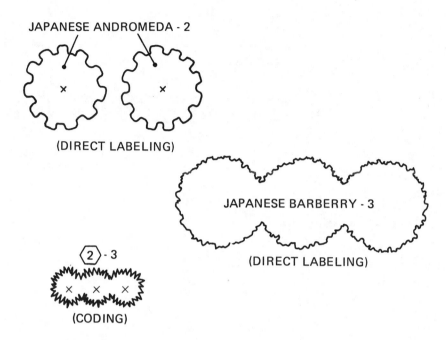

Fig. 11-2 Coding the labeling of plant symbols is less desirable than direct labeling but is sometimes necessary. Never use the techniques of direct labeling and coding on the same landscape plan.

PLANT SPECIES CODE

| #1 | COMMON LILAC | SYRINGA VULGARIS |
| #2 | CREEPING JUNIPER | JUNIPERUS HORIZONTALIS |

one label for several plants of the same species, the number of labels can be reduced and confusion avoided. An overabundance of labeling lines also can be confusing, and should be eliminated whenever possible.

Construction materials are labeled as plant materials are, but are seldom coded. While there is no need to indicate specific numbers (such as 30 bricks or 6 stones), the object to be constructed should be described according to such features as height, width, and color. For example, one labeling might read *basket weave fence – 5′ tall, redwood stained*. Another might read *concrete patio with natural redwood strips @ 2′ intervals*.

From the symbol labels, a tally can be made of how many of each species of plant are used in the total plan. Following the tally, the plant list is made. A *plant list* is an alphabetical listing of the botanical names of the plants planned for the landscape, their common names, and the total number used. Plant lists are

Fig. 11-4 A landscape plan complete with plant list.

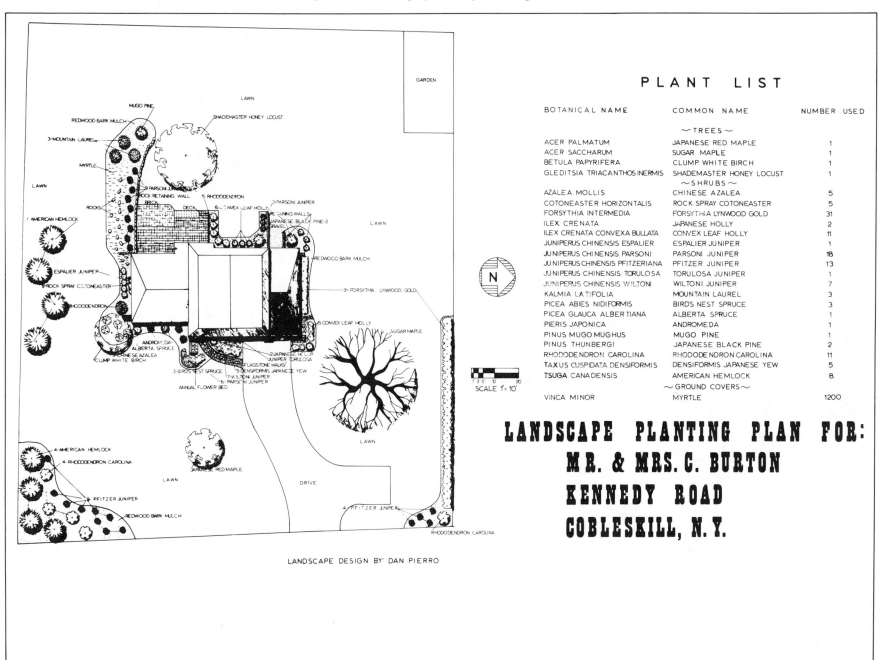

PLANT LIST

BOTANICAL NAME	COMMON NAME	NUMBER USED
~TREES~		
ACER PALMATUM	JAPANESE RED MAPLE	1
ACER SACCHARUM	SUGAR MAPLE	1
BETULA PAPYRIFERA	CLUMP WHITE BIRCH	1
GLEDITSIA TRIACANTHOS INERMIS	SHADEMASTER HONEY LOCUST	1
~SHRUBS~		
AZALEA MOLLIS	CHINESE AZALEA	5
COTONEASTER HORIZONTALIS	ROCK SPRAY COTONEASTER	5
FORSYTHIA INTERMEDIA	FORSYTHIA LYNWOOD GOLD	31
ILEX CRENATA	JAPANESE HOLLY	2
ILEX CRENATA CONVEXA BULLATA	CONVEX LEAF HOLLY	11
JUNIPERUS CHINENSIS ESPALIER	ESPALIER JUNIPER	1
JUNIPERUS CHINENSIS PARSONI	PARSONI JUNIPER	18
JUNIPERUS CHINENSIS PFITZERIANA	PFITZER JUNIPER	13
JUNIPERUS CHINENSIS TORULOSA	TORULOSA JUNIPER	1
JUNIPERUS CHINENSIS WILTONI	WILTONI JUNIPER	7
KALMIA LATIFOLIA	MOUNTAIN LAUREL	3
PICEA ABIES NIDIFORMIS	BIRD'S NEST SPRUCE	3
PICEA GLAUCA ALBERTIANA	ALBERTA SPRUCE	1
PIERIS JAPONICA	ANDROMEDA	1
PINUS MUGO MUGHUS	MUGO PINE	1
PINUS THUNBERGI	JAPANESE BLACK PINE	2
RHODODENDRON CAROLINA	RHODODENDRON CAROLINA	11
TAXUS CUSPIDATA DENSIFORMIS	DENSIFORMIS JAPANESE YEW	5
TSUGA CANADENSIS	AMERICAN HEMLOCK	8
~GROUND COVERS~		
VINCA MINOR	MYRTLE	1200

LANDSCAPE PLANTING PLAN FOR:
MR. & MRS. C. BURTON
KENNEDY ROAD
COBLESKILL, N.Y.

LANDSCAPE DESIGN BY: DAN PIERRO

SCALE 1" = 10'

N

used most frequently by the persons responsible for purchasing the plants.

On the completed final plan, the plant list is located opposite the design. Figure 11-4 illustrates the plant list and how it coordinates with the final plan.

Figure 11-4 also illustrates the other important landscape labels and their suggested positions. The directional arrow is a necessity. It provides the physical orientation for the property and permits the sun movement to be interpreted. The scale indicator is also vital. The largest lettering on the plan is reserved for the client's name. The designer's name is also included, but in smaller letters. Finally, the plan is bordered with a line running around the edge of the drawing. This border does the same thing that a frame does for a picture — it gives the entire plan a sense of unity.

Fig. 11-5 Vellum is placed over the original design for tracing.

Fig. 11-6 A diazo copier duplicates the landscape plan from vellum onto heavy paper in seconds.

Professional designers usually do their original design work on heavy, often gridded, paper. Then a sheet of thin, transparent tracing paper or *vellum* (a strong, cream-colored paper) is placed over it and the entire work traced, figure 11-5. The transparent tracing can then be copied on a duplicating device such as the diazo machine, figure 11-6. The diazo duplicator can be used to make unlimited copies of the plan on heavy paper or plastic film. Copies are then given to the landscape contractor who is responsible for the installation of the landscape.

ACHIEVEMENT REVIEW

A.   Select the best answer from the choices offered to answer each question.

1. What is the most desirable method of labeling plant symbols?

   a. near the symbol   b. by coding   c. on the symbol

2. What does the label *hemlock - 3* mean?

   a. A total of three hemlocks are referred to by the one label.
   b. The hemlocks are to be three feet apart.
   c. The hemlocks are to be three years old.

3. Which part of the plant list is alphabetized?

   a. The total number of plants used
   b. The common names of the plants
   c. The botanical names of the plants

4. Why is the directional arrow necessary on the final plan?

   a. It is ornate and attractive.
   b. It orients the property on the plan.
   c. It tells which way the wind blows across the property.

5. Which of the following is not a part of the final presentation plan labeling?

   a. plant list   b. designer's name   c. price list

SUGGESTED ACTIVITIES

1. Visit the office of a landscape architect. Ask to see examples of his or her work and request a demonstration of the use of the diazo duplicator.

2. Visit a drafting supplies store. Observe the many types of drawing instruments, papers, and furniture available to landscape designers.

3. Do a complete landscape design of your home, from the beginning drawings to the final tracing.

section 3

THE SELECTION AND USE OF PLANT MATERIALS IN THE LANDSCAPE

The starkness of desert plants makes their use a pleasure and challenge for the Southwestern landscaper.

Photo by Jack E. Ingels

Photo by Jack E. Ingels

Desert plants are used in a traditional grouping. Species are repeated to retain the desired simplicity. A planting such as this functions effectively as a divider of space, screen for partial privacy, or traffic director.

Existing trees provide the shaded setting for this attractive home. Foundation plantings are used to complement but not conceal the architecture.

Complementary shades of green contrast richly against the grey of the stone. Effects such as these require time to develop. Their appearance can only be partially predicted by a landscape designer.

Photo by Jack E. Ingels

Design and Photo by Goldberg and Rodler, Landscape Contractors, Huntington, N.Y.

This Family Living Area features a whirlpool. A sense of ceiling is provided, but without blocking the welcomed sunlight.

The brick walk, plant materials, and formal planters work together to lead the eye directly to the entrance of this home. Ground cover has been used in place of turf grass to form the floor of the outdoor room.

Photo by Jack E. Ingels

Photo by Jack E. Ingels

This planting of coleus and dusty miller creates a striking outdoor wall between a cafe and pedestrian walkway. Excellent color effects are attained by using plants which have colored foliage in addition to colored flowers.

Family Living Area pools require large expanses of solid surfacing around them to accommodate the many water- related activities of the family.

This patio is developed on multiple-levels to add interest and to create different areas of use.

Hanging gardens contribute aesthetically to the architectural design of this hotel in Kona, Hawaii.

Grading of this site permitted the development of a swimming pool and large patio on a steeply sloping property.

 # unit 12

TREES

OBJECTIVES

After studying this unit, the student will be able to

- select the proper tree to fill a certain role in a landscape.
- evaluate the possible strengths and weaknesses of a specific tree in a landscape.
- distinguish between bare-rooted, balled and burlapped, and containerized plants.
- explain the proper method of planting a tree.

As discussed in Section 2, one function of trees is to act as the ceiling of the outdoor living room. In that role, a tree may provide full or partial shade, while creating a feeling of intimacy or openness. To function correctly, trees must be selected with careful thought, not merely because the designer happens to like their appearance. The selection of a tree is determined by a combination of several factors:

- the height of the tree
- how low to the ground it branches
- the density of its foliage
- whether it is deciduous or evergreen
- seasonal color, foliage texture, and whether it bears flowers and/or fruit.
- hardiness
- ease of transplanting
- resistance to insects or diseases

TYPES OF TREES AND SHRUBS

Trees (and shrubs) are available and usable as native, exotic, or naturalized materials. *Native* plants are those which evolved through nature within a certain locale. Examples are the eastern white pine of the northeastern United States and the Douglas fir of the Pacific Northwest. *Exotic* plants are those that have been introduced to an area by individuals, not nature. For example, many of the junipers and yews used in landscaping in the United States actually evolved in China or Japan. However, they have adapted very well to America. *Naturalized* plants are those which were brought into an area as exotic, but have adapted so well that they have *escaped cultivation*. This means that they now occur commonly both in and out of planned landscape settings. The bird of paradise bush is such a plant. Native to South America, it grows like a native plant in the desert of the southwestern United States.

Exotics have certain advantages over many of the native plants. They often transplant more easily than the natives. Also, they often have fewer insect and disease problems. Such positive features are probably why over 60 percent of the trees and shrubs used in American landscaping are exotic in origin.

HOW TO SELECT A TREE

One important reason for selecting a particular tree is its hardiness rating. The *hardiness rating* determines whether or not the plant will survive the winter in the location desired. The United States Department of Agriculture has prepared a Hardiness Zone Map, figure 12-1, which shows the average annual minimum temperatures for all of the United States (except Alaska and Hawaii) and Canada. Notice on the map that the continent is divided into ten hardiness zones.

Each zone has an average annual minimum temperature variation of ten degrees Fahrenheit. As the hardiness zone number increases, so does the temperature minimum. For example, northern Kansas temperatures (Zone 5) drop to between −10° and −20° F in the winter, but southern Kansas (Zone 6) drops only to between

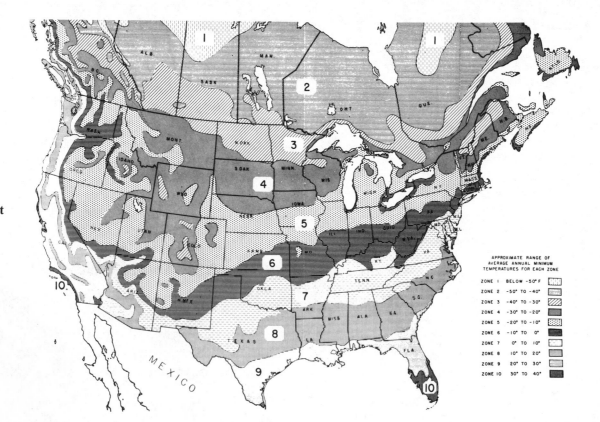

Fig. 12-1 U.S.D.A. plant hardiness zone map

APPROXIMATE RANGE OF AVERAGE ANNUAL MINIMUM TEMPERATURES FOR EACH ZONE

ZONE 1   BELOW −50° F
ZONE 2   −50° TO −40°
ZONE 3   −40° TO −30°
ZONE 4   −30° TO −20°
ZONE 5   −20° TO −10°
ZONE 6   −10° TO 0°
ZONE 7   0° TO 10°
ZONE 8   10° TO 20°
ZONE 9   20° TO 30°
ZONE 10 30° TO 40°

Fig. 12-2   Typical tree silhouettes, characteristics, and landscaping uses

Silhouette and Examples	Characteristics	Possible Landscape Uses	Silhouette and Examples	Characteristics	Possible Landscape Uses
wide-oval  Flowering crabapple Silk tree Cockspur hawthorn Flowering dogwood	• spreads to be much wider than it is tall • often a small tree • horizontal branching pattern • branches low to the ground	• focal point plant • works well to frame and screen • can be grouped with spreading shrubs beneath	**round**  Shinyleaf magnolia Cornelian cherry dogwood American yellow wood Norway maple	• width and height are nearly equal at maturity • usually dense foliage • if the tree is large, a heavy shade is cast	• lawn trees • mass well to create grove effect • larger growing species may be used for street plantings • smaller growing species can be pruned and used as patio trees
vase-shaped  American elm	• high, wide-spreading branches • majestic appearance • usually gives excellent shade • an uncommon tree shape	• excellent street trees • allows human activities underneath • frames structures • used above large shrubs or small trees • *note:* the American elm is easily killed by Dutch elm disease; this limits its use	**columnar**  Columnar Norway maple Columnar Chinese juniper Fastigiata European birch	• somewhat rigid in appearance • much taller than wide • branching strongly vertical	• useful in formal settings • accent plant • group with less formal shrubs to soften its appearance • frames views and structures
pyramidal  Pines Fir Spruce Hemlock Filbert Sweetgum Pin oak Sprenger magnolia	• pyramidal evergreen trees are geometric in early years • pyramidal deciduous trees are less geometric • pyramidal shape is less noticeable as the trees mature	• accent plant • large, high-branching trees allow human activity beneath • save older trees for their irregular shapes • *note:* avoid planting large trees near small buildings	**weeping**  Weeping willow Weeping hemlock Weeping cherry Weeping beech	• very graceful appearance • branching to the ground • easily attracts the eye • grass or other plants cannot be grown beneath them	• focal point plant • screens • attractive lawn trees • *note:* avoid grouping with other plants

acidity or alkalinity) is extreme, only a few shrubs or trees can be grown unless the soil is properly conditioned. To grow well, azaleas and rhododendrons require soil which is somewhat acidic and which contains large amounts of rich organic material. Likewise, yews should not be grown in poorly drained soil, since their roots cannot tolerate standing water. The soil in some areas, such as beneath black walnut trees, will not support the growth of any shrubs at all. (The walnut roots secrete a substance injurious to most plants that might be growing nearby.)

An increasingly popular reason for selecting certain shrubs is to provide a *habitat* (living and growing area) or food for birds and other small wildlife so that they are encouraged to remain in a particular area. For example, certain shrubs are ideal nesting sites due to their branching habits. Others produce tasty fruit which attracts birds, chipmunks, and rabbits. Most state colleges of agriculture can provide students with the names of shrubs in their particular state that are used for this purpose.

TRANSPLANTING SHRUBS

Successful transplanting of shrubs requires a procedure similar to the one described for trees in Unit 12. The important points are highlighted here.

A. Purchase the shrubs as bare rooted, balled and burlapped, or container grown. The transplanting season, temperatures in the area, and budget determine which type is specified by the landscape designer.

B. Prepare the planting site properly. Develop a rich, loose, loamy soil with the pH properly balanced. Make the planting hole at least 50 percent larger than the ball or root system it is to receive. Avoid the application of fertilizer to evergreens for the first year after transplanting. Use nonburning organic fertilizer if necessary for deciduous shrubs.

C. Tamp the soil down around the shrubs when backfilling. Build a ring of soil around the shrub to catch and hold rainwater. Use of an antitranspirant is advised with shrubs as well as trees.

D. Keep the shrub watered deeply and regularly for the first year. The use of an organic mulch around the shrub aids in moisture retention.

THE USE OF MULCHES

A *mulch* is a material placed on top of soil to

- aid in water retention.
- prevent soil temperature fluctuations.
- discourage weed growth.
- improve landscape appearance.

Mulches may be either *organic* (consisting of modified plant or animal materials) or *inorganic* (consisting of nonliving materials), figure 13-2. Organic mulches are more varied than inorganic mulches and are more commonly used.

When mulching shrubs (and trees), the mulch should be applied 3 to 4 inches deep for maximum effectiveness. A very shallow layer of mulch does not discourage weed seed germination, offer much water retention, or prevent constant changes in the soil

Fig. 13-2 Comparison of organic and inorganic mulches

ORGANIC MULCHES (peat moss, wood chips, shredded bark, chipped corncobs, pine needles)	INORGANIC MULCHES (marble chips, crushed stone, black plastic)
• reduce soil moisture loss • often contribute slightly to soil nutrition • may alter soil pH • are not a mowing hazard if kicked into the lawn • may be flammable when too dry • may temporarily reduce the nitrogen content of the soil	• also reduce moisture loss • do not improve soil nutrition • seldom alter pH • are a hazard if thrown by a mower blade • are inflammable • have no effect upon the nitrogen content of the soil

temperature. Repeated freezing and thawing of the soil around the plant's base can damage the bark and allow the entrance of disease or insects.

If a more shallow layer of mulch is desired, weed control and water retention can still be accomplished by spreading black plastic around the plant first and then adding enough mulch to weight down the plastic. It is essential that the plastic be black to prevent sunlight from penetrating and promoting weed growth. This mulching technique works well on flat land. However, it is not advised for use on slopes because rainwater tends to wash the mulch off the slick plastic. It must also be used cautiously on heavy clay soils; in these cases, the plastic may collect and hold too much water and drown the shrub or tree.

FERTILIZING SHRUBS

The nutrients found in soil are used by plants as they grow. Often, it is necessary to replace those nutrients by the application of chemical fertilizers. The fertilizers most commonly used by landscapers are those needed for maintenance of the garden. They are called *complete fertilizers,* which means that they contain the three primary nutrients needed in the greatest amounts by growing plants: nitrogen, phosphorus, and potassium.

For shrubs growing in cultivated beds, fertilizer may be applied in late March or early April. Depending upon the needs of the particular plants involved, each 100 square feet of bed area receives between 1 and 3 pounds of a low analysis, complete fertilizer (one which contains less than 30 percent total nitrogen, phosphorus, and potassium). The fertilizer is distributed uniformly on the soil beneath the shrubs. Care should be taken to place most of the fertilizer beneath the outer edge of the shrub rather than around the base. This is because the special roots which absorb the nutrients are centralized there. The fertilizer should not be allowed to touch the shrub's foliage, or foliar burn may result. *Foliar burn,* as described here, is a reaction of the leaf tissue to the harsh fertilizer chemicals. The tissue dies where it is touched by the fertilizer.

If the soil is dry, the fertilizer can be worked into the soil by use of a hoe. If the weather has not been abnormally dry, the fertilizer can be left untilled since the next rainfall will wash it into the soil.

All fertilization of shrubs and trees should be completed by July 1st. Application of the nutrients after that can cause *succulent* (fleshy) growth too late into the summer, making the plant more likely to receive injuries due to harsh winter weather.

THE USE OF SHRUBS IN LANDSCAPE DESIGN

Traffic control comprises a major use of the shrub planting as an outdoor wall. Figure 13-3 illustrates a shrub wall used to control pedestrian traffic. Figure 13-4 shows a similar use but displays the results of poor positioning of the shrub.

Fig. 13-3 This hedge, bordering the sidewalk, prevents pedestrians from cutting across the corner of the lawn.

Fig. 13-4 This hedge is set back too far from the sidewalk to function as an outdoor wall. Pedestrians have cut across the corner, destroying the lawn.

Fig. 13-5 The dark foliage of shrubs provides an excellent background against which to contrast sculpture.

Providing a background for flowers, sculpture, and fountains is another way in which shrubs are used for design purposes, figure 13-5. Most flowers and other enrichment items are displayed more attractively when contrasted against the rich green of a shrub mass.

Softening harsh building lines is a common function of shrub plantings, figure 13-6. The improvement in the overall appearance of the house is apparent.

In reducing wind velocity, shrubs are usually more desirable than a solid fence next to a patio. If the designer wishes to slow a strong

prevailing wind to a soft breeze, large shrubs can be effective, figure 13-7. A solid structure such as a fence or wall blocks the wind, but may not reduce its velocity unless it is designed with openings.

The fragrance of blooming shrubs such as roses, honeysuckle, and lilacs is sometimes a desirable landscape quality. When used near patios or under bedroom or kitchen windows, an aromatic shrub can add appeal to a home. Some landscape designers keep a list of fragrant trees and shrubs that are hardy in their area.

Screening is an important function where privacy or blockage of a view is desired by the designer. To screen properly, the shrubs

Fig. 13-6 Shrubs can be used to soften harsh building lines.

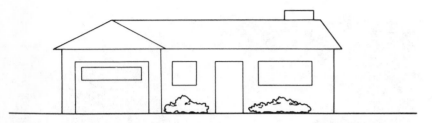

WITHOUT CORNER SHRUBS, THE LINES OF THE HOUSE SEEM STARK AND HARSH.

WITH CORNER SHRUBS, THE CORNERS ARE SOFTER AND LESS APPARENT.

Fig. 13-7 Shrubs can reduce the velocity of wind, turning a gust into a breeze.

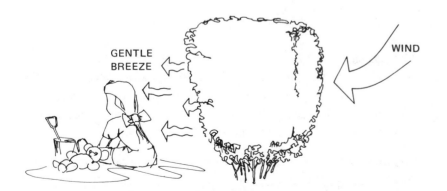

GENTLE BREEZE

WIND

must be dense and spaced closely enough that no one can see through them. Also, the height of the shrubs must be high as the eye level of the tallest viewer (around 6 feet). Figure 13-8 shows some of the instances in which shrub masses (often in combination with trees) are used to create visual screens.

The use of shrubs as a focal point and accent to the landscape was described earlier in the text. Because of their bright colors and various shapes, foliage textures, and branching patterns, many

Fig. 13-8 Various ways to screen with shrubs.

WITH FENCING. . .
TO PROVIDE SCREENING AND SECURITY

WITH TREES. . .
TO SCREEN AN UNSIGHTLY AREA

AS A HEDGE FOR PRIVACY

shrubs can be used alone to create points of interest. Thus, the shrub becomes an item of individual appreciation.

The following chart (A Guide to Landscape Shrubs) lists some of the more common landscape shrubs used in various parts of the country. This listing does not include every shrub available; rather, it is meant to act as an introduction. For a more complete chart, the student should make a separate list of the *flora* (plant life) which is not already included and which is common to his or her own area.

A GUIDE TO LANDSCAPE SHRUBS

Shrub	Evergreen*	Deciduous	Mature Height			Season of Bloom**			Light Tolerance			Good Fall Color	Zone of Hardiness	Comment
			3'-5'	5'-8'	8' and up	Early Spring	Late Spring	Early Fall	Sun	Semi-shade	Heavy Shade			
Almond, Flowering		X	X			X			X				4	very showy blooms
Azaleas														
Gable		X	X				X		X	X			6	requires an acidic soil
Hiryu	X			X			X			X			7	condition and often
Indica	X			X		X				X			8	iron chelate fertilizers
Kurume	X		X				X			X			7	
Mollis		X	X				X		X	X			6	
Torch		X	X				X			X		X	6	
Barberry														
Japanese		X		X			X		X	X		X	4	good plants for traffic
Mentor	semi			X			X		X	X			5	control; thorny
Red leaved		X		X			X		X	X			4	
Bayberry	semi				X				X	X			2	fragrant leaves and fruit good for seashore areas
Boxwood														
Common	X				X				X	X			5	prunes well; good for
Little leaf	X		X						X	X			5	formal hedges
Camellia	X			X		X				X	X		7	blooms from late October to April
Coralberry		X	X					X		X		X	2	good on banks for erosion control
Cotoneaster														
Cranberry		X	X				X		X	X		X	4	fall color comes from
Rockspray	semi		X				X		X	X		X	4	bright red fruit
Spreading		X		X			X		X	X		X	5	
Deutzia, slender		X	X				X		X				4	white flowers
Dogwood														red twig is very good
Cornelian cherry		X			X	X			X			X	4	for erosion control. All
Grey		X			X	X			X			X	4	dogwoods have good
Red twig		X			X	X			X			X	2	fall color
Firethorn														
Scarlet	semi			X			X		X			X	6	fall color comes from
Formosa	X				X		X		X			X	8	brightly colored fruit
Forsythia														
Early		X		X		X			X				4	bright yellow flowers
Lynwood		X			X	X			X				5	cascading branching
Showy border		X			X	X			X				5	patterns

Semi-evergreen indicates that the plants retain their leaves all year in warmer climates, but drop them during the winter in colder areas.

**Where no rating is given, flowers are either not produced or are not of importance.

A GUIDE TO LANDSCAPE SHRUBS

Shrub	Evergreen*	Deciduous	Mature Height 3'-5'	Mature Height 5'-8'	Mature Height 8' and up	Early Spring	Late Spring	Early Fall	Sun	Semi-shade	Heavy Shade	Good Fall Color	Zone of Hardiness	Comment
Gardenia	X		X				X	X		X			8	very fragrant flowers
Hibiscus														
Chinese	X				X	X			X				9	
Shrub althea		X			X			X	X				5	
Holly														
Chinese	X				X								7	fruit color is most
Inkberry	X				X								3	attractive in the fall
Japanese	X				X								6	
Honeysuckle														
Blue leaf		X			X		X		X	X			5	
Morrow		X		X			X		X	X			4	
Tatarian		X			X		X		X	X			3	
Hydrangea														
Hills of Snow		X	X				X		X				4	coarse-textured shrubs
Oak leaf		X		X				X	X			X	5	
Pee gee		X			X			X	X			X	4	
Jasmine														
Common white	semi				X		X		X				7	
Italian	X				X		X		X				8	
Primrose	semi				X		X		X				8	
Juniper														
Andorra	X		X						X				2	grows well in hot, dry
Hetz	X			X					X				4	soil
Japanese garden	X		X						X				5	
Savin	X			X					X				4	
Pfitzer	X				X				X				4	
Lilac		X			X		X		X				3	large, fragrant flowers
Mahonia														
Leatherleaf	X				X	X				X			6	holly-like foliage
Oregon grape	X		X			X				X			5	bluish, grapelike fruit
Mock orange		X			X	X	X		X				4	creamy white fragrant flower
Nandina	X			X				X	X	X		X	7	very attractive in flower and fruit stage
Ninebark		X			X		X		X			X	2	
Pieris (Andromeda)														
Japanese	X			X		X			X	X			5	
Mountain	X			X		X				X			4	
Pine, Mugo	X			X					X	X			2	
Poinsettia	X				X			late fall	X				9	long-lasting blooms
Pomegranate		X			X		X		X	X		X	8	colorful in both spring and fall
Potentilla (Cinquefoil)		X	X				X	X	X				2	produces yellow flower all summer

Semi-evergreen indicates that the plants retain their leaves all year in warmer climates, but drop them during the winter in colder areas.

**Where no rating is given, flowers are either not produced or are not of importance.

A GUIDE TO LANDSCAPE SHRUBS

Shrub	Evergreen*	Deciduous	Mature Height 3'-5'	Mature Height 5'-8'	Mature Height 8' and up	Season of Bloom** Early Spring	Season of Bloom** Late Spring	Season of Bloom** Early Fall	Light Tolerance Sun	Light Tolerance Semi-shade	Light Tolerance Heavy Shade	Good Fall Color	Zone of Hardiness	Comment
Privet														prunes well
Amur		X			X		X		X	X			3	popular hedge plants
California	semi				X		X		X	X			5	
Glossy	X				X			X	X				7	
Regal		X		X			X		X	X			3	
Quince, Flowering														densely branched,
Common		X		X		X			X	X			4	thorned plants
Japanese		X	X			X			X	X			4	good for traffic control
Rhododendron														
Carolina	X			X			X						5	
Catawba	X			X			X						4	
Rosebay	X				X		X						3	
Rose, Hybrid tea		X	X				X	X	X				varies	very diversified group of plants / special culture required
Spirea														
Anthony Waterer		X	X				X		X	X			5	very attractive when
Bridal wreath		X			X		X		X	X		X	4	blooming
Billiard		X		X			X		X	X			4	most are resistant to
Frobel		X	X				X		X	X			5	disease and insect pests
Thunberg		X	X			X			X	X		X	4	
Vanhoutte		X		X			X		X	X		X	4	
Viburnum														attractive spring flowers
Arrowwood		X			X		X		X	X		X	2	good fall foliage color
Black haw		X			X		X		X	X		X	3	many are good as wild-
Cranberrybush		X			X		X		X	X		X	2	life food
Doublefile		X			X		X		X	X		X	4	
Fragrant		X			X		X		X	X		X	5	
Japanese snowball		X			X		X		X	X		X	4	
Leatherleaf	X				X		X		X	X			5	
Sandankwa	X			X			X		X	X			9	
Weigela		X			X		X		X				5	blooms late
Winged Euonymus		X		X					X			X	3	spectacular crimson fall color
Wintercreeper	X		X							X		X	5	
Yew														excellent for founda-
Spreading Anglo-														tion plantings and
Japanese	X				X				X	X			4	hedges
Upright Anglo-Japanese	X				X				X	X			4	prunes well
Spreading Japanese	X				X				X	X			4	long lived
Upright Japanese	X				X				X	X			4	will not tolerate poorly
English	X				X				X	X			6	drained soil
Canada	X		X							X			2	

Semi-evergreen indicates that the plants retain their leaves all year in warmer climates, but drop them during the winter in colder areas.

**Where no rating is given, flowers are either not produced or are not of importance.

PRACTICE EXERCISES

A. Collect the fruit of ten or more different shrubs. Place each type in old plates or pie tins and set them outside. Observe which shrub produces fruit most attractive to the wildlife in your area. Which fruit is of no interest?

B. During the winter months, cut branches from early flowering shrubs. Bring them into the classroom and place in vases of water. Change the water every day. In this way, the shapes, sizes, and colors of the flowers can be observed before spring arrives. Which shrubs produce flowers before leaves? Which produce the leaves first?

C. Practice mixing good planting soil in the classroom. Collect soil from the home or school yard. Have a quantity of peat moss and sand also available. In a separate container, mix equal amounts of the three ingredients. Moisten the mix *slightly* and roll some between your fingers. If it sticks like modeling clay, more sand or peat is needed. If it crumbles, more soil is needed. When it rolls, but cracks slightly, the mixture is a *loam* and suitable for planting.

D. To demonstrate how mulch aids in water retention, fill two deep glass jars with equal amounts of dry soil. Leave about 4 inches unfilled. Slowly add enough water to thoroughly moisten the soil, without leaving water standing in the jar. Fill one jar to the top with a moistened mulch. Leave the other jar unmulched. Observe the two soils for the next week. Which soil dries out first?

E. Using drawing tools, design a corner planting with shrubs which would remain hardy in zone 5. Select the shrubs from "A Guide to Landscape Shrubs."

F. Design a line planting of shrubs suitable for a semishaded location in Zone 6.

G. Visit a nursery and compare the wide selection of shrubs available in your area.

ACHIEVEMENT REVIEW

A. Select the best answer(s) from the choices offered to answer each question.

1. Which two items are Latin botanical names of plants?

   a. Chinese juniper   c. Carolina rhododendron
   b. *Juniperus chinensis*   d. *Rhododendron caroliniana*

2. Why must a landscaper know both the common and the botanical name of plants?

   a. Clients are impressed by the use of botanical names.
   b. Some common names are localized and only the botanical name is reliable.

c. Some plants have only common names.

d. Clients recognize only botanical names.

3. In what landscape role do rigid, geometric shrubs function best?

a. as accent or specimen plants

b. as softeners of harsh building lines

c. on corners of buildings

d. as wildlife attractants

4. In what landscape role do soft, loose shrubs function best?

a. as accent or specimen plants

b. as softeners of harsh building lines

c. at the incurve of corner plantings

d. as windbreaks

5. Which of the following plants is not likely to be the same genus as all of the others?

a. grey dogwood

b. red twig dogwood

c. blue spruce

d. flowering dogwood

e. yellow-osier dogwood

6. Which two of the following plants are most closely related?

a. *Quercus rubra*

b. *Acer rubrum*

c. *Quercus borealis*

d. *Symphoricarpos albus*

7. How large should the planting hole be when transplanting shrubs?

a. slightly smaller than the root system or soil ball being set into it.

b. the same size as the soil ball or root system being set into it.

c. at least 50 percent larger than the root system or soil ball being set into it.

d. The size of the hole is not important.

8. What technique used in planting trees is not needed in transplanting shrubs?

a. soil conditioning

b. watering

c. mulching

d. staking

9. Which of the following are reasons to use a mulch around newly planted shrubs?

a. It holds moisture in the soil.

b. It reduces weed growth.

c. It reduces soil freezing and thawing.

d. all of these reasons

10. What role do fertilizers play in good shrub care?

a. They replace lost soil nutrients.

b. They reduce weed growth.

c. They prevent water loss.

d. They prevent windburn.

11. When should shrubs be fertilized?

a. middle of summer

b. July and August

c. late March to early April

d. whenever it is convenient

unit 14

GROUND COVERS AND VINES

OBJECTIVES

After studying this unit, the student will be able to

- distinguish ground covers and vines from other landscape plants.
- explain ways in which ground covers and vines are used to solve special landscape problems.
- list the methods for the installation and maintenance of ground cover and vine plantings.

The term *ground cover* is applied to small plants (less than 18 inches tall) which cover the ground in place of turf. The width of the plant does not matter; the definition is based upon the height and the use of the plant. There is a great variety of plants regarded as ground covers. Some such plants are small shrubs. Others are vinelike with long, trailing growth habits. Ground covers can be needled or broad-leaved evergreens, or deciduous. They can be *herbaceous* (nonwoody plants which die back every fall) or *woody* (leaving dormant stems and branches aboveground all winter). Some ground covers ease the maintenance requirements of gardens; others increase maintenance needs. Some ground covers yield attractive flowers; others do not.

Ground covers are generally regarded as special purpose plants, since many times they are used where other types of plants are not workable. Certainly turf grass is the most common material for covering the land's surface (although it is not classified as a ground cover). There are some conditions, however, under which grass does not grow well, such as places in which the soil is too dry, too acidic or alkaline, too steeply sloped, or too shaded. In such places, ground covers may be a better choice than grass. Figures 14-1 and 14-2 show successful ground cover plantings serving in functional

Fig. 14-1 Ground cover holds this steep bank and provides a lawn that does not require mowing.

Fig. 14-3 Ground cover in a raised bed provides an aesthetic quality while allowing water to reach tree roots.

Fig. 14-2 This ground cover, which fills the space between the building and sidewalk, grows equally well in shaded and unshaded areas.

roles. In figure 14-3, ground cover can be seen in an additional role, that of a beautifying design element.

HARDINESS OF GROUND COVERS

Although there are hardiness zone ratings for ground covers, they are often unreliable. The lack of reliability is due to the fact that ground covers grow so close to the soil. Soil holds both high and low winter temperatures longer than the atmosphere. As a result, there can be frequent thawing and refreezing of the ground during the winter. Repeated thawing and freezing can be very harmful to ground cover plants in that it *heaves* them to the surface of the soil, exposing their root systems to cold air and drying. Therefore, some ground covers may have a hardiness zone rating which implies that they will survive in an area, when they will not.

In other parts of the country, a heavy snow blanket is predictable throughout the winter. This blanket serves to insulate the ground and keep it frozen all winter. Without the freezing and thawing, the plants do not heave. The snow also protects the plants from the drying winds. Thus, some ground covers survive in regions colder than their hardiness zone rating suggests. Should there be a winter without the usual heavy snowfall, many ground cover plantings might die.

INSTALLING AND MAINTAINING GROUND COVERS

Soil Preparation

The soil for ground covers must have the ability to retain moisture. It should not be poorly drained, however. Since soil varies greatly throughout the country, the landscaper must analyze the needs of each particular soil at the planting site.

If the soil repels or fails to hold natural moisture, large quantities of organic material (peat moss, leaf mold, manure) should be worked into it. If the soil is compacted or heavy with clay, the addition of organic matter and sand may be helpful.

Adjustment of Soil pH

With over 200 species of ground covers from which to select, there is no one pH suitable for them all. The landscaper must know the pH which is appropriate for the plants being installed. Likewise, the existing pH of the soil must be determined. A simple pH test kit can be of great help in this.

If it is necessary to lower the pH of the soil, the addition of aluminum sulfate is recommended. About 2 to 4 pounds per hundred square feet is a good rate with which to begin. The actual rate depends upon the particular need at the time. The planting site should be thoroughly watered immediately after application of the aluminum sulfate.

The pH of soil can be raised by adding lime. It is best if the lime is applied in the form of dolomitic limestone. Depending upon how much adjustment is needed, from 10 to 25 pounds per hundred square feet are recommended.

Fertilizer Requirements

Most ground covers grow best if given some type of fertilization. Those growing near trees and shrubs are especially in need of it, since the roots of the larger plants compete with the ground covers for available nutrients. Cornell University recommends between 2 and 4 pounds of a 5 percent nitrogen fertilizer per hundred square feet. A 5 percent nitrogen fertilizer is one having an analysis such as 5-8-7 or 5-10-5.

The best time of year to apply fertilizer to ground covers is in the early spring when the most rapid growth occurs. If there has been any winter injury to the ground covers, the nutrients help the plants recover. The fertilizer should be watered in right away to prevent the foliage from being burned.

Spacing Ground Covers

If ground covers are used extensively in a garden plan, they can easily become the most costly of all plantings. Therefore, their spacing becomes important when the landscaper has a limited budget. The more closely the ground cover is planted, the more quickly a solid cover is formed. It also requires weeding over fewer years since the shade of a solid planting of ground cover prevents the growth of many weeds. The major drawback to close spacing at the time of planting is cost.

A planting of vining ground covers set 15 to 18 inches apart may take three or four years to fill in solidly. While the initial cost is less, there is much more weeding by hand or with costly herbicides needed over a longer period of time.

For small, vinelike ground covers (myrtle, Baltic English ivy, pachysandra), the most common spacings are 6, 12, or 18 inches. At a spacing of 6 inches, the planting requires approximately 6 plants per square foot. At a spacing of 12 inches, only 2 1/2 plants per square foot are required, while an 18-inch spacing needs only one plant per square foot. The differences in the cost of installation are quickly apparent.

If the ground cover plants are wide spreading but low growing shrubs (such as andorra juniper or rockspray cotoneaster), spacing is calculated in feet rather than inches. Regardless of the size of the plants, the planting arrangement is important. Greater coverage with fewer plants results if the plants are placed in a staggered fashion, figure 14-4.

Watering and Mulching Ground Covers

Ground covers must be encouraged to develop their roots deep into the soil as quickly as possible. In this way, they are able to survive warm, dry summer periods more easily. It is better to give ground cover plantings deep, thorough waterings at infrequent intervals than to give them frequent shallow waterings.

Mulches are sometimes used to conserve soil moisture in ground cover plantings. However, the type of material and the time it is done are important. Light, porous materials make the best mulches for ground covers. Materials such as peat moss, corncobs, and peanut hulls are excellent. *Caution:* Hay and straw should be avoided at all times because they contain seeds which can become weeding nightmares.

The initial mulching is done after the ground has frozen for the winter. This insulates the soil against thawing and refreezing. Thereafter, the mulching can be renewed at the same time each year. In areas where the ground does not freeze, the mulch is applied in early spring, prior to weed germination.

VINES

A *vine* is a plant which has a tendency to climb either naturally or when given proper support. Although it has height and width, it lacks the fullness that is normally associated with shrubs. Many ground covers can also be used as vines if they are given vertical support. Figures 14-5 and 14-6 illustrate uses of vines in the landscape.

Aside from their different shape and the different roles they play in the landscape, vines generally are planted and maintained as any other shrub. Some vines require special care, however, to protect them from the damaging effects of exposure to the sun and winter weather.

Fig. 14-4 Alternating the placement of ground covers fills space most efficiently.

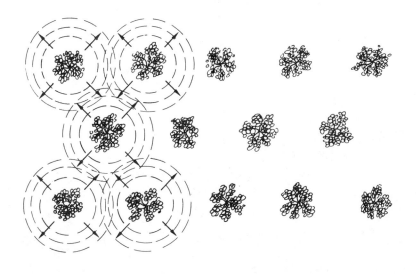

Fig. 14-5 The flowering wisteria provides large, grapelike clusters of blossoms.

Fig. 14-6 The owner of this Nantucket home is training vines to grow onto the roof for a vine-covered cottage effect.

Fig. 14-7 Vines growing against this tall building provide an attractive visual pattern.

Using Vines in the Landscape

Vines, the most flexible of all plants, serve many functions in the landscape. In the outdoor room, they create walls, ceilings, or floors, depending upon their support. They are also very effective in relieving the monotony of a large constructed wall, figure 14-7.

There are three ways in which vines climb, figure 14-8. Some vines climb by *twining* themselves around a trellis, fence, or another plant. Others produce fine *tendrils* that wrap around the supporting structure and allow the vine to climb. Still other vines produce *holdfasts* that permit the plant almost to glue itself to the support.

The landscaper should be aware of the climbing methods of vines before selecting them for use in the landscape. Otherwise, damage to the building could result. For example, vines which produce holdfasts can damage the walls of wooden buildings by pitting the surface and allowing moisture to seep through the wood.

Knowing the method by which vines climb allows the landscaper to

- select the proper species.
- construct the best support.
- avoid moisture damage to structures caused by a vine that is too dense or that will pit the surface of the supporting structure.
- avoid using a vine that is difficult to remove on a wall that requires frequent painting or maintenance.

The following tables list some of the more common ground covers and vines available for use in landscape design.

A GUIDE TO GROUND COVERS

Ground Cover	Evergreen	Deciduous	Height	Optimum Spacing	No. Needed to Plant 100 sq. ft.	Light Tolerance	Hardiness Zone Rating	Flower or Fruit Color and Time of Effectiveness
Ajuga or bugle		X	5"	6 inches	400	sun or shade	4	blue or white flowers in summer
Cotoneaster, creeping		X	12"	4 feet	10	sun	4	pink flowers, red fruit in summer and fall
Cotoneaster, rockspray	semi		18" plus	4 feet	10	sun	4	pink flowers, red fruit in summer and fall
Euonymus, big leaf wintercreeper	X		18" plus	3 feet	14	sun or shade	5	orange fruit in fall
Euonymus, purple leaf wintercreeper	X		18"	3 feet	14	sun or shade	5	not of significance
Honeysuckle, creeping		X	12"	3 feet	14	sun	5	pale yellow flowers in spring; red fruit in fall
Ivy, Baltic English	X		8"	18 inches	44	shade	4	none
Mondo	X		12"	10 inches	144	partial shade	8	white or pink flowers in spring
Myrtle or Periwinkle	X		8"	12 inches	92	shade	4	blue flowers in spring
Oyster Plant		X	12"	12 inches	92	sun or shade	9	not of significance
Pachysandra	X		12"	12 inches	92	shade	4	white flowers in spring
Sarcococca	X		tall — requires shearing	3 feet	14	sun or shade	7	white flowers and scarlet berries in fall
Wandering Jew	X		6"	12 inches	92	shade	9	red-purple flowers in spring and summer
Weeping lantana	X		18" plus	24 inches	25	sun	9	lavender flowers all year
Yellowroot		X	18" plus	18 inches	44	sun	5	brown-purple flowers in spring

A GUIDE TO VINES

Vine	Broad-leaved Evergreen	Deciduous	Height	Clinging	Twining or Tendrils	Light Tolerance	Hardiness Zone Rating	Flower or Fruit Color and Time of Effectiveness
Actinidia, bower		X	30'		X	full sun or semishade	4	white flowers in spring
Actinidia, Chinese		X	30'		X	full sun or semishade	7	insignificant
Akebia, fiveleaf	semi		35'		X	full sun or semishade	4	purple flowers in spring
Ampelopsis, porcelain		X	20'		X	semishade	4	multicolored fruit in fall
Bignonia (or crossvine)	X		60'		X	full sun or semishade	6	orange-red flowers in spring
Bittersweet, American		X	20'		X	sun or semishade	2	yellow and red fruit in fall and winter
Boston ivy		X	60'	X		sun or shade	4	insignificant
Bougainvillea	X		20'	X		full sun	7	multicolored in summer
Clematis		X	3' to 25'*		X	full sun or semishade	4 to 7*	many colors of flowers in late spring
Euonymus, evergreen bittersweet	X		25'	X		sun or shade	5	yellow and red fruit in fall and winter
Fig, creeping	X		40'	X		sun or shade	9	insignificant
Honeysuckle, trumpet		X	50'		X	full sun or semishade	3	orange flowers in summer; red fruit in fall
Hydrangea, climbing		X	75'	X		full sun or semishade	4	white flowers in summer
Ivy, English	X		70'	X		semishade	5	insignificant
Kudzu vine		X	60'		X	sun or shade	6	insignificant
Monks hood vine		X	20'		X	semishade	4	yellow-orange fruit in fall
Rambling roses		X	10' to 20'		support needed	sun	5	flowers of many colors in spring and summer
Trumpet vine		X	30'	X		sun	4	orange flowers in summer
Virginia creeper		X	50'	X		sun or shade	3	insignificant
Woodbine, Chinese		X	50'		X	shade	5	yellow flowers in summer; red fruit in fall

*Dependent upon the actual species selected

Fig. 14-8 How vines climb

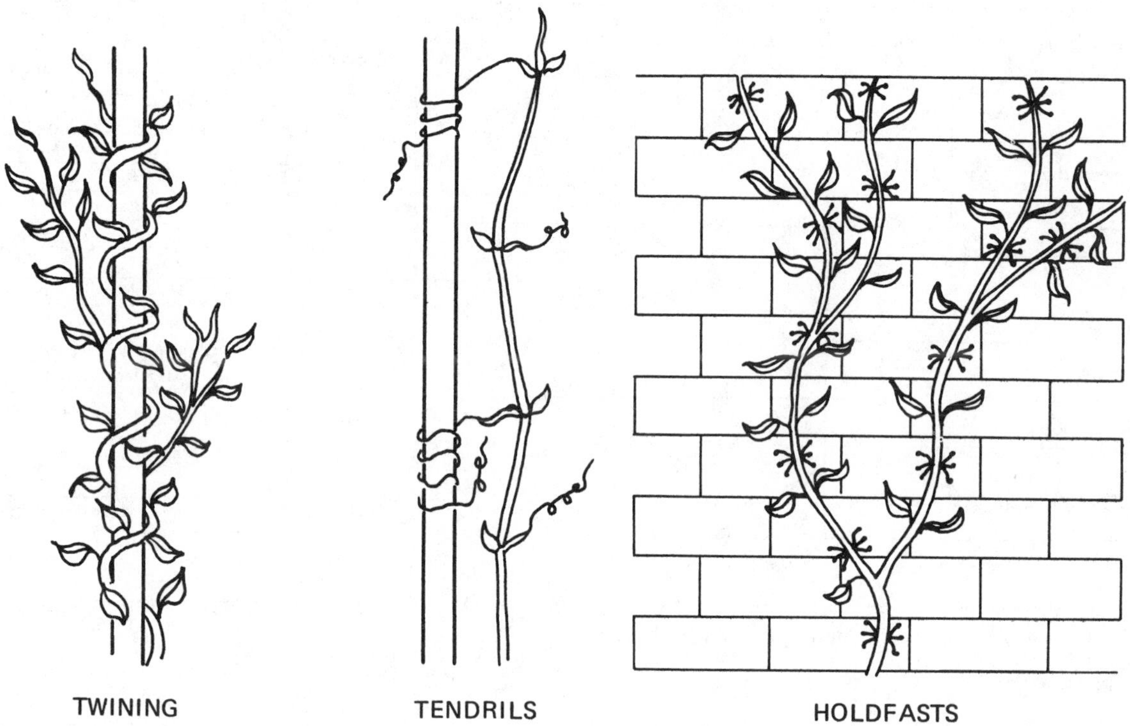

TWINING   TENDRILS   HOLDFASTS

ACHIEVEMENT REVIEW

A. Indicate whether the following characteristics describe ground covers (G), vines (V), both ground covers and vines (B), or neither ground covers nor vines (N) by marking with the appropriate symbol.

may be deciduous or evergreen

are usually 18 inches or less in height

have great height and width, but lack fullness

often produce colorful flowers and fruit

can serve as ceilings in the outdoor room if properly supported

have the most unreliable hardiness zone ratings

most easily harmed by repeated freezing and thawing of the soil

serve no useful role in the landscape

B. Describe the proper method of planting and maintaining ground covers. Include the following items.

- soil preparation
- adjustment of the soil pH
- fertilization
- spacing
- watering and mulching

C. Explain the different ways in which vines climb.

SUGGESTED ACTIVITIES

1. Propagate several ground covers. To accomplish this, take some cuttings (4 to 5 inches in length) from ground cover plantings near the school. Select herbaceous types, since they root faster than woody types. Remove the lower 2 1/3 to 3 inches of leaves and place in flower pots or greenhouse flats of pasteurized sand. Be certain that the container drains well. Keep the sand moist. If a greenhouse and humidity chamber are not available, enclose the pots in a bell jar or within a sealed plastic bag to maintain a high humidity. Check the cuttings for roots in two or three weeks. Transplant into soil after rooting.

2. Find some areas where ground covers are needed. Visit a nearby park or woodlot. Seek out the spots where grass is not growing well; determine why it is not. Which areas could be planted to bear ground covers and which could not? Why?

3. Locate vines near the school. Study the ways in which they climb. Look for any signs of injury to the supportive structure. Rate the vines on the basis of foliage density, seasonal color, and how easy removal of the vine would be if necessary. Determine the function of the vine in the landscape.

unit 15

FLOWERS

OBJECTIVES

After studying this unit, the student will be able to

- differentiate among annual, perennial, and biennial flowers.
- list the characteristics of hardy and tender bulbs.
- explain the difference between flower borders and flower beds.
- design a flower planting.
- install and maintain a flower planting.

Flowers are valuable elements of any landscape design, and must be chosen with great care. Flowers have a quality which can easily make them the focal point in a landscape, often in conflict with more important objects. Their value lies in the many colors, fragrances, and seasonal tones that they naturally possess.

Imagine a spring without daffodils and tulips or a summer without geraniums and petunias. Picture an autumn without chrysanthemums or a Christmas season without poinsettias. The flowers of our landscapes are strong reflections of the four seasons. The wise landscape designer uses flowers often but with care.

ANNUALS

An *annual* flower is one which completes its life cycle in one year. That is, it goes from seed to blossom in a single growing season and dies as winter approaches. Generally, annuals are most commonly used in summer landscapes. They bloom throughout the months of June, July, August, and September when the days are long and warm. They offer color, especially in northern regions, when many bulbous perennials are past blooming. There are hundreds of annuals commonly used throughout the country. Some examples of well-known annuals are the petunia, marigold, salvia, and zinnia.

Landscapers obtain annuals by two different methods. In one method, they are directly seeded into the ground after the danger

Fig. 15-1 Packaged seeds are an easy and inexpensive way to start a flower garden.

Fig. 15-2 Bedding plants are available from the greenhouse ready to plant in the garden. They give color faster than the direct-seeding method, but cost more.

of frost is past. Packages of annual seeds, figure 15-1, are available at most garden centers in the spring. Seeds may also be obtained from mail-order supply houses.

Direct seeding of annuals (the placement of seeds in the ground) is the least expensive way to place flowers in the landscape. The major limitations of direct seeding are: (1) the young plants usually require thinning; (2) the plants require more time than the other method to reach blooming age; and (3) it is difficult to create definite patterns in the flower planting.

The other method used to obtain annuals is from *bedding plants*. These are plants which were started in a greenhouse and are already partially grown at the time they are set into the garden, figure 15-2. Very often, the plants are grown in a pressed peat moss container which can be planted directly into the ground.

This creates no disturbance for the annual's root system; thus, the flower planting is well established from the very beginning.

Bedding plants are more expensive than seeded annuals. However, there is no need for thinning, and definite patterns can be easily created.

PERENNIALS

A *perennial* flower is one which does not die at the end of its first growing season. While it may become dormant as cold weather approaches, it lives to bloom again the following year. (When a plant is *dormant,* it is experiencing a period of rest in which it continues to live, but has little or no growth.) Most perennials live at least three or four years; many live for much longer.

Nearly all early spring flowers are perennials. Some grow from bulbs; others do not. Many special autumn flowers are also perennials. There are numerous summer perennials which act with annuals to add color to rock gardens and border plantings. Some examples of perennials are the hyacinth, iris, daffodil, tulip, poppy, phlox, gladiolus, dahlia, and mum.

Some perennials are available in seed form. However, the majority appear on the market as bedding plants or reproductive structures such as bulbs. Since they do not die at the end of the growing season, most perennials reproduce themselves and may eventually cover a larger area of the garden than was originally intended. This tendency to propagate may be a side benefit or a maintenance nuisance, depending upon the situation.

Bulbous Perennials

The very popular bulb comprises a large number of perennials. Bulbs survive the winter as dormant fleshy storage structures known

to botanists as tubers, corms, rhizomes, tuberous roots, and true bulbs. In the landscape trade, they are usually simply called *bulbs*.

Most bulb perennials bloom only once, be it in the spring, summer, or fall. There are a few exceptions, however; these may bloom repeatedly. Bulbs may be classified as hardy or tender.

Hardy bulbs are perennials which are able to survive the winter outside and therefore do not require removal from the soil in the autumn. The only time they must be moved is when they are being thinned. Hardy bulbs usually bloom in the spring. Examples are the hyacinth, iris, daffodil, and tulip.

Tender bulbs are perennials which cannot survive the winter and must be taken up each fall and set out each spring after the frost is gone. These bulbs usually bloom during the summer months. Examples are the canna, gladiolus, caladium, and tuberous begonia.

Figure 15-3 illustrates the autumn installation of a sizeable bulb planting around a large piece of outdoor sculpture. Figure 15-4 shows the planting as it appeared the following spring.

Fig. 15-3 Autumn: installing a large bulb planting in front of an outdoor sculpture.

Fig. 15-4 Spring: the bulb planting in bloom.

BIENNIALS

A smaller group of flowers known as *biennials* complete their life cycle in two years. They produce only leaves during their first year of growth and flower the second year. After they have bloomed, they die. Biennials include the English daisy, foxglove, Japanese primrose, and pansy.

FLOWER BEDS AND FLOWER BORDERS

A *flower bed* is a freestanding planting made entirely of flowers. It does not share the site with shrubs or other plants. As figures 15-5 and 15-6 illustrate, flower beds are effective as focal points where sidewalks or streets intersect. They also work well in open lawn areas where they do not conflict with more important focal points and where the lawn is not used for activities that could be damaging to the beds.

Flower beds should never be planted in the public area of a residential landscape. When this is done, the beds attract more attention than the entry to the home, which is the most important part of any public area.

Flower beds must be designed so that they can be viewed from all sides, and must be planted accordingly. Because of this requirement, flower beds contain no woody plants to provide a backdrop for the blossoms. This has given the flower bed the reputation of being the most difficult flower planting to design. Probably for this reason, it is used much less often than the flower border.

The *flower border* is a planting which is placed in front of a larger planting of woody shrubs. The foliage of the shrubbery provides a background to set off the colors of the blossoms, figure 15-7. Since the flower border can only be viewed from one side, it is more easily controlled by the designer. Figure 15-8 illustrates the difference between the viewing perspectives of beds and borders.

Modern landscapers are more likely to use flowers in borders than beds for two reasons: they are easier to design, and the strong visual attraction of the flowers is more easily controlled.

Fig. 15-5 This freestanding flower bed acts as a traffic divider while adding beauty to the area.

Fig. 15-6 Walks on all sides create various points from which this flower bed may be viewed. Despite its size, only two species were used (salvia and dusty miller).

Fig. 15-7 The flower border has a background, as opposed to the flower bed. Here, a vine-covered wall creates a rich contrast for blossoms.

Fig. 15-8 Flower beds are viewed from all sides and have no background foliage. Flower borders have a more limited viewing point and have background foliage.

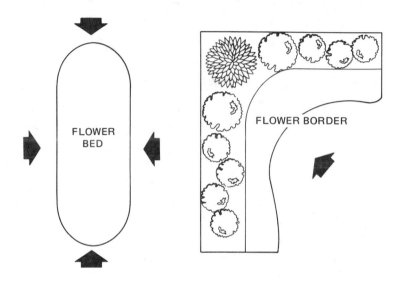

DESIGNING FLOWER PLANTINGS

The placement of flowers in a bed or border arrangement is similar to shrub placement in most ways. Flowers are arranged with the tallest to the rear and/or to the center. They are grouped in masses rather than placed as individuals. Flowers, like all other plant materials in the landscape, are placed according to the principles of design.

Additional factors to remember when designing flower plantings are:

- All colors must blend together attractively; orange and red flowers can exist comfortably in the same planting if they do not bloom at the same time.

- Only those flowers blooming at any one time will be noticed.
- Dark, vibrant colors attract the eye strongly and should be used sparingly.
- Pale, pastel colors attract the eye less and can be used in greater quantities.
- Rigid, sharply formed blossoms attract the eye and should be used at the center of the planting.
- Round, loosely formed blossoms attract the eye less and should be used in the foreground and at the ends of the planting.

If an all-seasons flower planting is being designed, species must be selected and arranged so that there will always be a vibrantly colored, rigidly formed flower blooming in the center of the planting. The species will change with each season, but the forms should be similar. Flowers that are shorter, looser, and lighter in

Fig. 15-9 One example of how to arrange flower shapes, colors, and sizes in the flower border

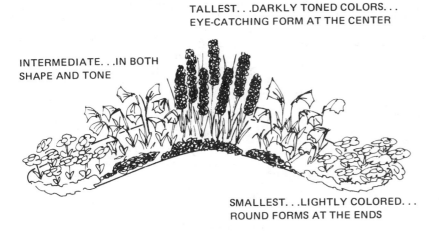

TALLEST...DARKLY TONED COLORS...
EYE-CATCHING FORM AT THE CENTER

INTERMEDIATE...IN BOTH
SHAPE AND TONE

SMALLEST...LIGHTLY COLORED...
ROUND FORMS AT THE ENDS

Fig. 15-10 Bedding plants are placed in a staggered arrangement for maximum coverage. Peat pots (on left) do not have to be removed before planting.

color should be selected to complement the central form. They too must be chosen carefully so that one or more species is always in bloom.

INSTALLING AND MAINTAINING FLOWER PLANTINGS

Flowers require a rich, loamy soil that holds moisture yet drains well. Being shallow-rooted, flowers do not do well in soil which allows water to run off rather than soak in, or which has so little organic matter that moisture is not retained. When preparing soil for flower seeding or transplanting, the ground should be cultivated and conditioned to a depth of at least 12 inches. As with trees, shrubs, and other plants, the amount of sand or peat moss to be added is determined by the particular soil conditions at the site.

Directions for installation are given on seed packages for the direct-seeding method. These directions must be followed carefully. Later, the young seedlings may be thinned and transplanted.

With the bedding plant method, the flowers are separated individually and usually spaced between 8 and 12 inches apart.

Placement is staggered as with ground covers to maximize coverage. Each young plant is set into the ground so that the soil level is about 1 inch higher than it was around the plants in the greenhouse, figure 15-10. This extra coverage is especially important if the plants are slightly straggly when transplanted. At the time each plant is set into the ground, a cup of water (perhaps containing a weak fertilizer solution) poured into the hole before backfilling helps reduce transplant shock for the young plant.

Bulbs must be set into the ground at the proper time of year if they are to bloom on schedule. Hardy bulbs must be planted about October in order to bloom the following spring, since they need the cold winter weather to form their flower buds. Tender bulbs can be set out only after the ground has thawed and there is

A GUIDE TO BULBOUS PERENNIALS

Bulb	Flowering Time	Time of Planting	Hardiness	Height	Planting Depth to Top of Bulb	Spacing
Amaryllis	summer	spring	tender	15″ or more	set top to ground level	12 inches
Anemone	summer	spring	semihardy*	7″ to 14″	2 inches	12 inches
Bulbous iris	summer	spring/fall	semihardy*	15″ or more	2 inches	12 inches
Caladium	summer (foliage color only)	spring	tender	7″ to 14″	2 to 3 inches in northern sections; 1 inch in southern sections	8 inches for mass effects
Calla lily	summer	spring	tender	7″ to 14″	set top at ground level	12 inches
Canna	summer	spring	tender	36″ or more	3 to 4 inches	18 to 24 inches
Crocus	spring	fall	hardy	2″ to 6″	3 inches	2 to 4 inches
Daffodil	spring	fall	hardy	7″ to 12″	4 to 5 inches	6 to 8 inches
Dahlia	summer	spring	tender	20″ or more	5 to 6 inches	24 to 36 inches
Gladiolus	summer	spring	tender	24″ or more	3 to 4 inches	6 to 8 inches
Grape hyacinth	spring	fall	hardy	1″ to 4″	3 inches	2 to 4 inches
Hyacinth	spring	fall	hardy	7″ to 12″	4 to 6 inches	6 to 8 inches
Lily	summer	spring/fall	hardy	15″ or more	6 inches	12 inches
Ranunculus	summer	spring	semihardy*	7″ to 14″	2 inches	12 inches
Snowdrop	spring	fall	hardy	1″ to 6″	3 inches	2 to 4 inches
Summer hyacinth	summer	spring	tender	1″ to 6″	4 inches	6 to 8 inches
Tuberous begonia	summer	spring	tender	7″ to 12″	set just below soil surface	6 to 8 inches
Tulip	spring	fall	hardy	7″ to 20″	4 to 5 inches	6 to 8 inches

*Semihardy bulbs are regarded as tender in northern sections and hardy in southern sections.

no danger of frost. In the fall, tender bulbs must be dug up, the soil shaken off, dried, and stored in a cool, dry, dark area until the next summer. It is helpful to dust the bulbs with a fungicide/insecticide mix while they are out of the ground. This helps to protect them from disease and insect problems. All injured or infected bulbs should be discarded rather than stored and replanted.

The chart above provides the beginning landscaper with basic information concerning common bulbous perennials.

Following their installation, all flowers should be thoroughly watered and mulched. An organic mulch applied to a depth of about 3 inches helps to conserve soil moisture and reduce weeds.

Maintenance of flower plantings varies with the type. Annuals can be fertilized in midsummer with a low analysis fertilizer to keep them lush and healthy. Bulbs should be fertilized immediately after flowering with a high phosphorus fertilizer such as bone meal. Nonbulbous perennials grow best if fertilized in the early spring.

Summer or fall fertilization can actually harm the plants by keeping them too succulent as winter approaches.

All flowers should be watered frequently and deeply during dry periods. Their shallow roots quickly react to drought conditions.

After the bulbs have ceased to flower, their foliage must be allowed to grow until it dies back naturally. It is during this post-flowering period that the food needed by the bulbs for the next season's growth is being produced.

About halfway through the summer, annuals frequently begin to look straggly and set seed, usually reducing their flower show.

In these cases, it is common practice to give the flowers a severe pruning. With bushy flowers such as the petunia, a pair of grass clippers is used to cut the plants back. With more stalky annuals, a severe pinching has the same effect. The plants may be unattractive at first, but new shoots and fresh flowers form within two weeks or so. The planting looks fresh and new and carries its bright colors on into the fall season.

As winter approaches, annuals should be cut off at ground level or removed entirely from the planting bed. Perennials can be cut back and mulched. Tender bulbs should be dug up and stored.

ACHIEVEMENT REVIEW

A. Indicate whether the following statements apply to annuals (A) or perennials (P).

The plant does not die at the end of the growing season.

These plants complete their life cycle in a single year.

These plants are most commonly used during the warm summer months.

Some forms of these plants are bulbous.

Nearly all of our spring flowers are of this type.

B. Indicate whether the following statements apply to hardy bulbs (H), tender bulbs (T), or both (B).

The bulbs can survive the winter without being moved inside.

The bulbs usually bloom only once each season.

These bulbs are planted in the spring and bloom during the summer.

These bulbs are planted in the fall and bloom the following spring.

The foliage of these bulbs should not be cut back until it has turned brown naturally.

C.   Indicate whether the following statements apply to flower beds, flower borders, or both.

The flowers in this planting are freestanding with no background shrubs.

In this planting, the tallest center flowers should have the brightest color.

This planting is only viewed from one side, making it easier to design.

This planting is mulched to help retain moisture during the season.

The flowers in this planting grow best in a rich, loamy soil.

SUGGESTED ACTIVITIES

1.   Write to seed companies and request catalogs. The gardening section of your newspaper may have some addresses of suppliers. Posting the pictures around the room will help you become familiar with common flowers in the area.

2.   With drafting tools, design a flower bed or border. It should cover an area of at least 40 square feet. Design it so that there are plants in bloom from early spring to late fall.

3.   Start annual flowers from seed. The seed can be purchased at a local garden center, supermarket, or hardware store. Plant them in greenhouse flats or flowerpots and place in a greenhouse or near a sunny window. When the weather is warm enough, set them outside around the school building. As an added activity, design a planting for the building first, using the flowers being grown in class.

4.   Practice blending flower colors. Cut out patches of colored paper to represent different types and quantities of flowers. Arrange them so that the brightest, most attracting colors are used where the attraction is desired. Pastel colors should be used in greater quantities and placed away from the central color. Follow the suggestions given in the text for arranging flower colors.

unit 16

THE ARID LANDSCAPE

OBJECTIVES

After studying this unit, the student will be able to

- name the regions of the United States where arid landscapes are common.
- explain how landscapes in arid regions differ from others in the country.
- characterize the soils of arid regions.
- describe methods for retaining water in arid soils.
- identify plants that are suitable for arid landscapes.
- explain the methods of transplanting and maintaining cactus plants.

THE ARID LANDSCAPE DEFINED

An *arid* landscape is one whose plants receive little usable water. The water available to the plants may be limited by quantity or quality, by such environmental circumstances as drying winds, or by a combination of factors.

Arid Regions of the United States

Arid landscapes are common to the states of Arizona, Colorado, New Mexico, and Texas, and parts of California, Nevada, Oklahoma, and Utah.

THE SOUTHWEST

In contrast to the soft, lush green, temperate and subtropical regions of the United States, the Southwest stands apart. This region is distinctive not only by its climate, but also by its culture, flora, and fauna. In the past, the Southwest was usually thought of in terms of its native Indians, Mexican and cowboy legacies, Spanish architecture, towering land forms, vast deserts, and strange plant materials. It seemed to be a world apart from the rest of the country and, for most of its history, was home to a very small percent of the population. There were few planned landscapes, and even fewer gardeners who were interested in learning about an area so alien to them.

All of this changed with the onset of the energy crisis, and the resultant shifting of the population. The Sun Belt has the fastest growing population in the country. New homes and businesses are being built throughout the region as individuals and industry seek reduced heating costs and milder winters. The full impact on the Southwest has yet to be seen. Only time will show the effects of the non-traditional and imported architecture, lifestyles, and plant material brought in by people from northern and eastern states.

The Changing Landscape

Changes in the landscape have occurred as the Southwest became home to people transplanted from New England, the Atlantic Northeast, the Midwest, and the Northwest. Some people have attempted to make the strange arid landscape closely resemble the more familiar outdoor rooms of the temperate states. In certain cities, such as Phoenix, large areas of the local flora have been replaced and transformed by out-of-state exotic materials. Bluegrass lawns grow on desert sand, and eastern shrubs bloom from planting beds filled with carefully blended soil mixtures. Those who appreciate the differences unique to each geographical region do not welcome this trend toward uniformity.

The professional landscaper seeking work in the Southwest will find four distinct differences between arid and temperate landscapes:

- *The soil quality is poor* throughout most of the Southwest.

- *Irrigation* is an absolute necessity throughout the year.

- *Altitude variations* result in extremely hot daytime temperatures and very cool nights.

- *High winds* dry out plants quickly and often damage them physically.

SOILS OF THE ARID REGION

Arid soils generally fall into three categories: pure sand or gypsum, adobe, and caliche.

Sand has almost no nutrient content and is without humus, the important organic component of soil. *Adobe* is a heavy clay-like soil, used by early residents to make building blocks. Adobe holds moisture better than sand but needs humus to lighten it and improve its aeration. *Caliche* soils are highly alkaline, due to an excessive lime content. They have a calcareous hardpan deposit near the surface that blocks drainage, making plant growth impossible.

Caliche soil is the most problematic soil a landscaper is likely to encounter. Heavy irrigation watering causes the alkaline lime layer to form near the soil surface. The hardpan deposits may lie right at the surface of the soil or at a depth varying from several inches to several feet below ground level. The deposits may occur as a granular accumulation or as a compact, impermeable concrete-like layer.

Caliche soils are not a problem in the deep soils of valleys. However, in locations outside of valleys, the land will neither drain naturally nor support healthy plant growth where unaltered caliche is present.

Generally, arid soils

- lack humus.

- require frequent irrigation.

- are nutritionally poor; nutrients are continually leached out by the irrigation water.

- are highly alkaline (pH of 7.5 to 8.5 and higher).

- are low in phosphate.

- lack iron or else contain it in a form unavailable to the plants.

- have a high soluble salt content resulting from alkaline irrigation waters, manures, and fertilizers that do not leach thoroughly.

Organic matter must be added to all southwestern soils to improve their physical structure. The organic matter improves the water retention capability of light sandy soils and breaks up heavy adobe soils. The only way to improve the drainage of caliche soil is to break through it and remove the impermeable layer. The excavated soil can be replaced with a soil mix that will support healthy plant growth.

WATER RETENTION

Water problems are common to landscape gardening throughout the Southwest, whether the landscape is on the valley slopes or the desert floor. An adequate supply of water is never available. What water there is may be so alkaline that it creates as many problems as it cures. Highly alkaline irrigation water can counteract an acidic soil created to support the growth of such plants as azaleas and rhododendrons. It can also contribute to the buildup of soluble salts in the soil. This results in increasingly weakened plant growth.

Nevertheless, irrigation watering is vital to the good health of arid landscapes. However, it is essential that *all* water be retained, whether it is applied naturally or by irrigation.

Methods of Retaining Water

The following methods are used to retain water and moisture in the arid landscape.

- All planting beds in the arid landscape should be recessed several inches below ground level to create a catch basin, figure 16-1. This method traps and holds any applied water, preventing loss through run-off.

Fig. 16-1 A recessed planting bed creates a catch basin for retaining moisture.

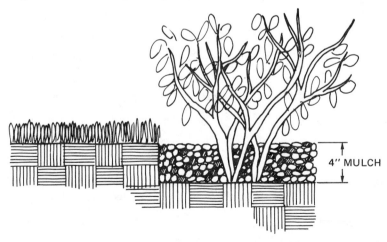

RECESSED PLANTING BED

4" MULCH

- Enough irrigation water should be applied to permeate several inches into the soil, avoiding excessive surface evaporation.

- Organic matter must be incorporated into the soil each year to improve the soil's water retention capacity.

- Organic mulches must be applied to a depth of at least four inches. The mulch slows moisture loss for the soil, and creates a cooler growth environment for the roots.

- Young trees and new transplants can lose enough water through their thin bark to suffer severe damage during a hot summer. For that reason, trunk wraps and whitewash paint should be applied to guard against sun scald. The same remedy is helpful to citrus trees and freshly exposed areas of recently pruned older trees.

THE SELECTION AND USE OF PLANTS IN THE ARID LANDSCAPE

As landscaping expands in the Southwest, certain questions come to mind. What plants are available and suitable for gardens in arid regions? What types of gardens are appropriate for the area? Can familiar temperate zone plants be expected to survive in arid gardens? Is the landscape profession in the Southwest totally different from that in other regions of the country? The answers to the questions are not clear-cut because the opinions of people vary, just as the many local climates and soils within the Southwest vary. No simple answers will suffice for a region of the country that includes such diverse areas as the Grand Canyon and Palm Springs.

Selecting Plants for the Arid Landscape

Landscapers familiar only with temperate zone plants often are dismayed when they first encounter the flora of arid regions. The succulents and cacti seem stark and strange to the eye. The trees they do recognize are generally regarded as weed trees in the eastern and midwestern states. The lush green cool-season grasses, such as bluegrasses and fescues, are missing. In their place are

A GUIDE TO SELECTED SOUTHWESTERN PLANTS

Plant	Growth Habit	Mature Height							Season of Bloom					Special Use in the Landscape
		1' or less	2'–5'	6'–9'	10'–15'	15'–30'	30'–50'	Over 50'	Early Spring	Late Spring	Summer	Fall	Winter	
Ash														
Arizona	T						X		NS					shade tree
Modesto	T						X		NS					shade tree
Citrus trees	T				X				varies with the variety					excellent for containers
Coral tree	T					X			X					brilliant flowers
Crabapple, flowering	T					X				X				specimen plant
Cypress, Arizona	T						X		NS					screens and windbreaks
Elderberry, desert	T					X			NS					screens and windbreaks
Elephant tree	T					X			NS					
Elm														
Chinese	T						X		NS					shade tree
Siberian	T						X		NS					windbreak
Eucalyptus	T						X		varies with the variety					many species prized for flower and/or foliage
Hackberry, netleaf	T						X		NS					shade tree
Honeylocust														
Shademaster	T					X			NS					good in dry, desert conditions
Sunburst	T					X			NS					
Thornless	T						X		NS					
Ironwood, desert	T		X							X				
Jujube, Chinese	T					X				X				specimen tree very salt tolerant
Locust														frequent pruning makes these attractive flowering trees
Black	T							X	X		X			
Idaho	T						X			X				
Pink flowering	T						X			X				
Magnolia, southern	T							X				X		lawn tree
Mesquite														shade trees and windbreaks
Honey	T					X			NS					
Screwbean	T					X			NS					
Mulberry, white	T							X	NS					shade tree

A GUIDE TO SELECTED SOUTHWESTERN PLANTS (continued)

Plant	Growth Habit	Mature Height							Season of Bloom					Special Use in the Landscape
		1' or less	2'-5'	6'-9'	10'-15'	15'-30'	30'-50'	Over 50'	Early Spring	Late Spring	Summer	Fall	Winter	
Olive, European	T					X			NS					good multi-stemmed tree
Pagoda tree, Japanese	T					X					X			lawn tree
Paloverde														specimen trees
Blue	T					X				X				
Little leaf	T					X				X				
Mexican	T					X				X				
Pine														grows well in poor soil specimen plant
Aleppo	T						X		NS					
Digger	T						X		NS					good in desert conditions
Italian stone	T							X	NS					good in planters; prune well
Japanese black	T					X								
Pinyon	T				X									multi-stemmed effects
Pistache, Chinese	T							X	NS					good patio tree good fall color
Poplar														narrow columnar form windbreaks
Balm-of-Gilead	T						X		NS					
Bolleana	T						X		NS					
Cottonwood	T							X	NS					
Lombardy	T							X	NS					
White	T						X		NS					
Silk tree	T						X				X			showy shade tree
Smoke tree	T				X					X				
Sycamore														excellent street trees
American	T						X		NS					
Arizona	T						X		NS					
California	T						X		NS					
Tamarisk														wind, drought, and salt resistant
Athel tree	T						X				X			
Salt cedar	T					X					X			
Umbrella tree, Texas	T						X			X				shade tree
Willow														
Babylon	T						X		NS					
Globe Navajo	T							X	NS					
Wisconsin	T						X		NS					

A GUIDE TO SELECTED SOUTHWESTERN PLANTS (continued)

Plant	Growth Habit	Mature Height							Season of Bloom					Special Use in the Landscape
		1' or less	2'–5'	6'–9'	10'–15'	15'–30'	30'–50'	Over 50'	Early Spring	Late Spring	Summer	Fall	Winter	
Zelkova, sawleaf	T						X		NS					windbreak
Abelia, glossy	S			X							X			
Apache plume	S		X						X					
Arborvitae, Oriental	S				X				NS					
Barberry														barrier plantings
Darwin	S			X						X				
Japanese	S		X						NS					
Bird of paradise	S			X							X	/		
Brittlebush	S		X							X				
Butterfly bush	S			X							X			vigorous growth
Cherry laurel, Carolina	S					X			X					screens and hedges
Cotoneaster, silverleaf	S			X						X				wind screen
Crape myrtle	S					X					X			very colorful
Creosote bush	S			X							X			screens and hedges
Firethorn, Laland	S			X					X					espaliers well
Hibiscus														
Perennial	S			X							X			
Rose of Sharon	S				X						X			
Holly														Wilson and Yaupon clip and shade well
Burford	S			X					NS					
Wilson	S			X					NS					
Yaupon	S					X			NS					
Hopbush	S			X					NS					screens
Jojoba	S		X						NS					hedges
Juniper														
Armstrong	S		X						NS					
Hollywood	S				X				NS					
Pfitzer	S			X					NS					
Lysiloma	S				X					X				good for transition between garden and natural landscape

A GUIDE TO SELECTED SOUTHWESTERN PLANTS (continued)

Plant	Growth Habit	Mature Height							Season of Bloom					Special Use in the Landscape
		1' or less	2'-5'	6'-9'	10'-15'	15'-30'	30'-50'	Over 50'	Early Spring	Late Spring	Summer	Fall	Winter	
Myrtle	S		X								X			prunes and shapes well
Ocotillo	S				X					X				specimen plant
Oleander	S				X						X	X		does well in heat and poor soil
Photina	S					X			X					screens
Privet														all species can be pruned to lower heights
California	S				X				X					
Glossy	S					X			X					
Japanese	S				X					X				
Texas	S			X						X				
Rose, floribunda	S		X							X				massing effects
Silverberry	S				X				NS					good for containers
Sugar bush	S				X				X					
Bougainvillea	V			X							X			very colorful
Ivy														
Algerian	G	X							NS					
Boston	V					X			NS					
Jasmine, star	V					X					X			very fragrant
Lavender cotton	G	X									X			effective as edging
Periwinkle	G	X								X				
Trumpet creeper	V					X						X		
Virginia creeper	V					X			NS					
Wisteria	V				X					X				may be trained as shrubs and weeping trees

T   Trees
S   Shrubs
V   Vines
G   Ground covers
NS Flowers are not showy.

coarser varieties, tall ornamental grasses, or just sand. Missing also are forested slopes for the landscape to reach toward, and rolling meadow views to see through windows in outdoor walls.

Once the differences are noted, the landscaper must look further to find comforting similarities. Deciduous and evergreen forms of plants can be found. The landscaper will find trees, shrubs, vines, ground covers, and flowers to work with. The landscaper will still find it necessary to deal with soft irregular plant forms and rigid geometric forms. (In this situation, the geometric, round barrel cactus and the globe arborvitae have something in common. Both are difficult to incorporate into a unified landscape design.) Likewise, the functions of the plants used in southwestern landscapes are similar to those of plants anywhere else in the country. A windbreak is just as important in New Mexico as it is in Kansas. A specimen plant fills the same focal-point role, whether it is a crape myrtle in Georgia or an unusually shaped black locust in Nevada. In short, any species of plant should be judged for its value and function in a particular landscape. One region's weed tree is another region's specimen plant; neither region should claim horticultural superiority.

The Guide to Selected Southwestern Plants in this unit summarizes some of the plants commonly used in landscaping the arid regions of the United States. It is intended not as a complete list, but as a starting point to use when selecting plants. More detailed and complete lists are available from books that deal specifically with southwestern gardening. The Cooperative Extension Service of individual southwestern states can also be of assistance.

Using Plants in the Arid Landscape

The plants in arid regions look quite different from those in other areas. It seems as though the gardens of the arid regions follow a different set of rules in their development. How else could such unusual designs result?

After careful analysis of well-designed southwestern landscapes, it can be seen that the principles of design are as valid in the arid garden as they are in temperate and subtropical gardens. These principles would be weak indeed if they did not apply to all situations. Attention still focuses on the entry to a building and selected points within other use areas. Balance is still important, with the informal, asymmetrical types used most often. Simplicity, which is inherent to the wide-open spaces that characterize the Southwestern landscape, is basic to the man-made landscape.

Proportion is another important consideration. Southwestern architecture, especially residential architecture, is mainly single story. Small trees are more suitable than tall ones to maintain the proper viewer perspective and size relationship between the buildings and the landscape. The principle of rhythm and line, while still valid, leans toward looser, more open space formations than in more populated regions of the country.

Some distinctions do exist. For instance, the space efficiency needed in an urban penthouse garden is seldom a factor in a southwestern garden. Also, arid climates do not have the abundance of moisture and pH neutral soils common to many American gardens which permit the growth of a wide variety of plants. Further, the sheer starkness of the southwestern plants, especially in the desert region, makes the viewer more aware of them individually than in temperate or subtropical gardens. In the latter regions, the massing of foliage is a common practice, and is easily done.

The most successful arid landscapes draw their inspiration and materials from the natural countryside around them. The use of native plant materials, stones, adobe bricks, gravels, and enrichment items that reflect the region's culture creates a garden that suits the locale. Conversely, the use of imported temperate zone plants and turf grasses creates a landscape that is at odds with its natural setting. An example of this is the landscape currently evolving in Phoenix.

Figures 16-2 through 16-6 show a variety of typical landscaping designs suitable for southwestern areas.

Whether the garden is inspired by the surrounding countryside or is an attempt to duplicate a spot in Connecticut, it should be informal in style. The typical southwestern home is of Spanish or Pueblo architecture. Recent variations in building trends have brought the western ranch and other modern styles into the southwestern

Fig. 16-2 This Arizona home is enframed in a balanced landscape. Attention focuses on the entry, as it should.

Fig. 16-3 The public area of this home is nicely designed. The number of species is limited, forms are massed, and the materials used are all native to the area.

Fig. 16-4 This landscape would benefit from more massing of plants, with less emphasis on individual plant forms.

Fig. 16-5 Massing of plants draws attention away from individual plant forms. These prickly pear cacti lend themselves well to such a use.

Fig. 16-6 Southwestern landscaping should reflect the native elements of plants, stones, sand and space, rather than introducing elements alien to the area.

Fig. 16-7 How a natural windbreak works

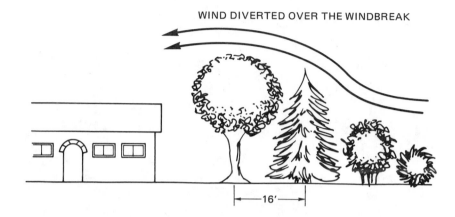

cities and suburbs as well. None of these styles lends itself well to formal, symmetrical landscape designs.

As the professional landscaper seeks to meet the needs of the southwestern client, the same concerns arise that are common to landscapers everywhere. How much privacy is desired? How much upkeep will the client accept? How can the climate be modified? Where is the nearest source of supply for materials? How can the indoors and outdoors be tied together most effectively?

Climate Modification

Satisfying the client's needs, while trying to achieve a sense of landscape unity without violating the integrity of the natural site is a complicated challenge. The first factor to consider is climate modification. Most homeowners want relief from the torrid summer sun and the forceful winter winds. Southern and western locations are

the best for the planting of small-to-medium-sized trees that will shade the house during the warmest hours of the day. The northern and eastern sides of the house offer the longest periods of shade. Therefore, these are the best sites for the development of family living area patios. This is one significant difference from landscaping in temperate regions, where the southern and western sides of a house are preferred for patio development.

Winter winds blow mostly from the west. They are best controlled by methods that diminish their force and attempt to divert it. Vertical louvered fencing and/or planted windbreaks are best for this purpose. The most effective natural windbreak combines both trees and shrubs. Four or five rows planted 16 feet apart, starting with shrubs on the windward side and building up to trees on the house side, work best. See figure 16-7.

The Importance of Natural Features

Whether treated as an on-site or off-site feature, the natural landscape must be respected. Successful southwestern gardens are those that improve upon the natural setting. The use of stones, repetition of earth colors, addition of cooling water features, and native materials all help to tie the man-made and natural landscapes more closely together. Views across the desert or toward a distant

Fig. 16-8 The foreground trees enframe the distant view.

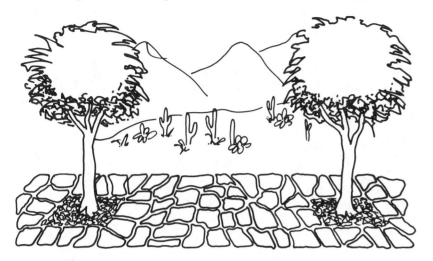

Fig. 16-9 The on-site feature should match the mood of the natural backdrop and be visually dominant.

mountain range should be enframed to become part of the landscape, figure 16-8. Where appropriate, the off-site natural landscape can be used as a backdrop for an on-site enrichment feature. In such a case, the on-site feature should match the mood of the natural backdrop and be visually dominant, figure 16-9.

Desert Gardens

With desert gardens, there is often a desire to develop the landscape with the native cacti and succulents. Not all desert plants are easily or freely transplanted, though. Some are quite difficult to transplant. Others are protected by law to safeguard them against destruction or theft. Many native plants grow quite slowly, and are endangered species in their natural range. Landscapers should check local laws before collecting from the wild. Many of the desired plants can be purchased from southwestern nurseries where they are grown for sale.

Where desert plants and nondesert plants are combined in the same landscape, it is best to group the species into plantings that have similar cultural requirements. For example, plants that cannot tolerate wind should be grouped against the northeast wall. Plants that require special soil mixtures should not be placed in

the same bed as cacti. Plants that need frequent watering should be placed close to the house; plants that need less water may be placed at more distant points.

PLANT INSTALLATION AND MAINTENANCE

The basic methods of plant installation for arid regions are similar to those described in Units 12 through 15. The soil must be conditioned to promote drainage and aeration. Large and frequent applications of humus-making organic material are necessary. Sand must be added to adobe soils. If caliche soil is involved, enough of the soil must be replaced to create a totally new root environment to sustain the plant for its entire lifetime.

Cactus Plants

Cactus plants are sufficiently different from other plants to warrant special mention as to how they can be transplanted successfully.

• Before transplanting, note and mark the north side of the cactus. It is essential that this side of the plant be reoriented to the north in its new location. Otherwise, the intense southern/

southwestern sun will harm the plant. The plant will have developed a thicker layer of protective tissue on its south side to withstand the heat.

- By trenching around the cactus, lift as much of the root system as possible.

- Brush as much soil as possible from the roots, then dust them with powdered sulfur.

- Place the cactus in a shaded, open area where air circulates freely. Allow damaged roots to heal for a week before replanting.

- Plant the cactus in dry, well-drained soil. Stake it if necessary.

- Water *after* new growth starts, usually in three or four weeks. Thereafter, water at monthly intervals.

General Guidelines

It is best to plant native desert species before the onset of summer heat. Early spring is a good time, and fall is even better. The soil must drain well, since the roots of most desert plants rot easily if grown in poorly drained soil. Planting during the cool seasons permits strong roots to develop before the summer arrives. This gives the plants a better chance to survive.

Watering depends upon the season and the species of plant. Cacti need water as often as once a week during their period of active growth in the spring. During the fall and winter, they need little or no water and should be allowed to go dormant. Succulents should be watered as soon as the soil surface dries out. It is recommended to water succulents several times weekly during the active season, and once every two weeks during the winter. Lawns must be deeply watered at least twice weekly. All plants should be hosed off periodically to wash away dust. Water must not be applied so late in the day that foliage is still moist in the cool night air.

Desert gardens require far less maintenance than do temperate and subtropical gardens. The dry climate discourages many of the insects and diseases that trouble gardens elsewhere.

Weeds are a maintenance nuisance in desert gardens, as they are everywhere. In small areas, the weeds can be removed by hand. In large areas, weeds can be controlled by chemical herbicides, but there is a possible *hazard* in using chemicals in arid regions. An unexpected rain shower can carry the herbicide from its area of application to nearby desirable trees and plants. For this reason, it is best to use weed killers that kill the weed upon contact and then quickly become inactive in the soil.

Fertilizing the southwestern garden is vital, since the soils are low in natural fertility. A soil test should be made to determine precisely what a local soil needs. For the region in general, phosphates and iron chelates or iron sulphate will need to be added. Organic fertilizers, such as sewage sludge, cottonseed meal, and animal manures are especially good for southwestern soils. Among chemical fertilizers, the 4-12-4 analysis is used most often. As with plants throughout the country, fertilizers should be applied preceding the periods of most rapid growth. They should not be applied later in the growing season when they would encourage soft growth, which would be damaged by winter weather.

ACHIEVEMENT REVIEW

A. Which of these states are part of the arid southwestern region of the United States?

1. Wyoming
2. Colorado
3. Texas
4. Utah
5. Oregon
6. Nevada
7. North Dakota
8. Oklahoma
9. Idaho
10. California
11. Arizona
12. South Dakota
13. New Mexico
14. Nebraska

B. List four features of southwestern landscapes that make them different from landscapes in other parts of the United States.

C. Name the three distinct soil types of the arid region.

D. Indicate if the following characteristics apply best to adobe soils (A), caliche soils (C), pure sand soils (S), or all soils of the Southwest (AS).

1. lacking in humus
2. very heavy soil; bricklike
3. low in phosphate content
4. impermeable hardpan deposit near surface
5. alkaline pH
6. the poorest nutrient content
7. low in iron content
8. high soluble salt content
9. holds some moisture
10. allows no drainage

E. List five methods of retaining water in southwestern plants and plantings.

F. Select the word or phrase that best completes the following statements.

1. The best locations for shade trees in an arid region are off the _____ sides of the house.
    a. north and south   c. east and west
    b. south and west   d. north and east

2. Family living area patios are most desirable on the _____ sides of the house in the Southwest.
    a. north and south   c. east and west
    b. south and west   d. north and east

3. Southwestern winter winds blow mainly from the _____.
    a. north   c. east
    b. south   d. west

4. Windbreaks for a southwestern property are most effective when located off the _____ side of the house.
    a. north   c. east
    b. south   d. west

5. When the natural landscape is used as a backdrop, the enrichment item used against it should match in style and be _____.
   a. dominant   c. inconspicuous
   b. recessive   d. color coordinated

6. Plant groupings in arid landscapes should be based upon a similarity of _____ requirements.
   a. client   c. nutrient
   b. designer   d. cultural

7. Cacti develop a thickened layer of protective tissue on their _____ side, and should be oriented carefully when transplanting.
   a. north   c. east
   b. south   d. west

8. Before replanting, cacti to be transplanted should be _____.
   a. kept moistened for a week
   b. pruned back
   c. balled and burlapped
   d. shaded for a week in an open area so their roots can heal

9. The best season for planting southwestern gardens is _____.
   a. spring   c. fall
   b. summer   d. winter

10. The best herbicides for arid region plantings are those that _____.
    a. kill upon contact and remain active in the soil
    b. kill upon contact and then quickly become inactive in the soil

SUGGESTED ACTIVITIES

1. Collect samples of gypsum, adobe, and caliche soils. Try not to break the calcareous, hardpan layer of the caliche. Compare the soils for water permeability by applying equal measured amounts to each sample and measuring the amount of water collected from the base of each sample after a fixed time period.

2. Using the Guide to Selected Southwestern Plants, identify those plants on the list that are found in your region of the Southwest. Add to the list using a catalog from a nearby nursery.

3. Convert planting designs found in this and other texts that use temperate zone plants into similar arrangements using southwestern plants.

4. In two boxes of soil, each 12 inches deep, place thermometers at depths of 2, 4, 6, and 8 inches. Leave one box unmulched. Cover the other with an organic mulch to a depth of 4 inches. Keep both boxes equally watered and set out in the full sun. Record the temperatures of both boxes at all levels. The cooling influence of the mulch should be apparent. How does the mulch affect water retention? How does it affect soil temperature? How does the soil temperature influence root growth?

section 4

USING CONSTRUCTION MATERIALS IN THE LANDSCAPE

 # unit 17

ENCLOSURE MATERIALS

OBJECTIVES

After studying this unit, the student will be able to

- list five functions of landscape enclosure.
- describe five types of materials used for constructed landscape enclosures.
- explain the need for and means of releasing water pressure from behind solid enclosure materials.

FUNCTIONS OF LANDSCAPE ENCLOSURE

The walls of the outdoor room serve to enclose (surround or separate) a section of property which is to act as an outdoor living area. *Enclosure materials* form these walls.

There are several functions of landscape enclosure:

- To define the shape and limits of the landscape
- To control circulation patterns within the landscape
- To provide various degrees of protection and/or privacy
- To serve other engineering needs, such as retaining steep slopes and raising planting beds
- To modify the climate by creating sheltered areas where plants may grow, or by diverting or reducing the wind's force to make the area more comfortable for occupants

As discussed previously, enclosure is sometimes accomplished with plant materials. Many other types of enclosures are man-made; these may be used alone or in combination with plant materials.

The type of constructed enclosure selected by the landscaper is usually determined by the function it is to perform. If its only purpose is *aesthetic* (to beautify the landscape and make it appealing to the senses), the most important consideration is how well the enclosure material coordinates with the total design.

> To achieve total unity of design, repeat colors and materials used in the building(s) and in the surfacing and furnishings of the garden in the enclosure materials.

CHOOSING MAN-MADE ENCLOSURES BY FUNCTION

Directing Circulation Patterns

If the purpose is to direct circulation patterns, a fence or wall may be used. Height is usually more important than density when traffic control is the concern. Therefore, open style fencing can be used if the viewer is to be able to see through the enclosure

structure, figure 17-1. The higher the enclosure is, the more effective it is in directing movement. A combination of an open style enclosure and plants can create an outdoor wall that is both functional and attractive, figure 17-2.

Protection

If protection is the objective, a solidly constructed fence or wall is a necessity. Open style fences, such as the chain link, allow the viewer to see through them while discouraging passage. Walls offer the greatest protection since they combine both density and strength. The height of the fence or wall chosen is determined by the type of protection it is to offer. To prevent children or small dogs from wandering into the street may require an enclosure only 3 feet high. To prevent intruders or large animals from entering, a wall 6 feet high or higher is required.

Fig. 17-1 Post and rail fencing can direct traffic patterns without blocking the view.

Fig. 17-2 The combination of constructed and natural enclosure materials is often more attractive and effective than either type would be if used alone.

Fig. 17-3 Privacy between two properties on the same level requires an enclosure of at least 6 feet in height.

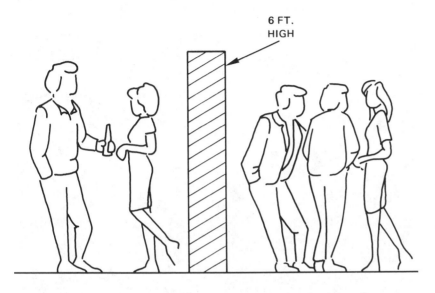

Fig. 17-4 The combination of a constructed enclosure and trees gives privacy when the viewer is above ground level.

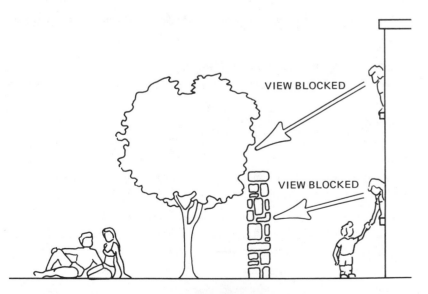

Privacy

For privacy control, the height and density of the enclosure material are again important. Where properties are small and family living areas are closely spaced, as in many older city neighborhoods, enclosure for privacy is a necessity. The enclosure must be solid where total privacy is needed, but a cramped feeling can result when a solid enclosure is used in a small space. Using a combination of plants and constructed enclosures helps to avoid this feeling.

The height of the enclosure necessary for privacy depends upon the location of the persons desiring the privacy and the vantage point of an unwelcome viewer. If a homeowner wishes to be screened from a next-door neighbor when both are in their family living areas at ground level, the enclosure should be about 6 feet high, figure 17-3. If the neighbor looks out a second-floor window onto the patio, the vantage point has been elevated, giving the neighbor an unobstructed view. This requires that the height of the enclosure increase also. Constructed enclosures cannot be much higher than 8 feet without becoming too imposing or threatening in appearance. In addition, there may be local laws limiting fence height. Therefore, trees are frequently used to block views in certain cases, figure 17-4.

When determining the enclosure height needed for privacy, remember:

- When both the viewer and the view are parallel to the ground, privacy can be obtained by a 6-foot enclosure at any point along the viewing line. It does not matter if the land is flat or sloped as long as the line of sight parallels it, figure 17-5(A).

- When the viewer and the view do not parallel the ground, the enclosure height necessary for privacy increases as the enclosure approaches the viewer, figure 17-5(B). The height decreases as it approaches the person desiring the privacy.

Engineering Uses

There are many engineering uses of constructed enclosures. The raised planting bed, for example, is one method of providing grade variation without bulldozing ground. The raised bed also

Fig. 17-5 The differences in the height of the enclosure necessary for privacy when the line of sight is and is not parallel to the ground.

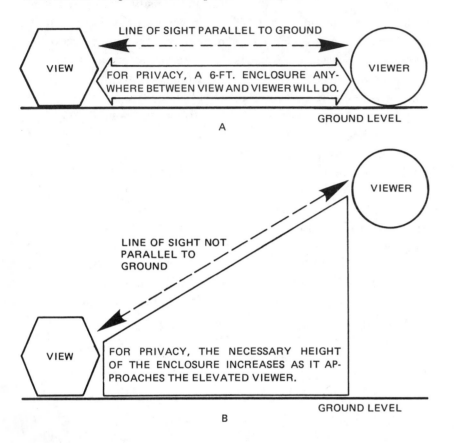

Fig. 17-6 The raised planting bed adds an interesting effect to this flat roof-top garden. It also offers protection by keeping people away from the edge of the roof.

permits plant growth where the existing soil or drainage conditions are unsatisfactory, figure 17-6. An associated benefit of raised planting beds is that they can also help control traffic or function as seatwalls (walls with flat, smooth tops suitable for use as seats).

Constructed enclosures in the form of *retaining walls* can transform slopes that are too steep for human use into smaller, usable areas. They may also be used to protect steep banks from erosion. For example, the bank can be turned into earthen steps which greatly reduce the downward rush of water, figure 17-7.

When constructed enclosures are used in engineering roles, it is important that the landscaper choose very strong materials. In

some cases, tons of soil may be held back by a retaining wall. The landscaper must also carefully specify the materials to be used, such as stone, poured concrete, or brick.

Releasing Pressure Behind Retaining Walls. Each construction material has its own special requirements for installation. Some require a footing or internal steel bracing; others need wooden forms during pouring.

One concern with which all landscapers must deal when retaining walls are part of the design is the release of pressure caused by water which can build up behind the walls. Construction of retaining walls often alters the normal drainage pattern of the land. Water which once flowed easily down a bank may have no way of getting past a wall which blocks the direction of flow. If not released, the water may eventually channel under the wall and weaken or burst the wall.

Fig. 17-7 What was once a steep slope has been converted into a series of broad, level planting strips connected by steps. The brick retaining walls provide both strength and beauty.

Fig. 17-8 Weep holes

Fig. 17-9 A constructed enclosure creates a sheltered area for the growth of marginally hardy plants.

Weep holes, figure 17-8, are the means by which water is allowed to escape from behind retaining walls. Built at the bottom of walls, they provide the necessary drain for water as it seeps down through the soil.

Climate Modification

Where climate modification is the objective of the constructed enclosure, either fences or walls may be used. It may also be desirable to combine a constructed element with plants for even better control or a more attractive appearance. A solid wall or fence built across the path of a prevailing breeze may be able to eliminate the breeze as a factor of the landscape near the wall. If a gentler breeze is desired rather than total stillness, an open style wall or fence is used. This slows down the wind's velocity without totally eliminating air movement.

In areas of borderline hardiness, certain species will survive the winter if they are given a protected planting site. A con-structed enclosure provides sheltered areas for marginally hardy plants in figure 17-9.

ACHIEVEMENT REVIEW

A. List five functions of constructed enclosures.

B. Select the best answer from the choices offered to complete each statement.

1. Enclosure can be accomplished with constructed materials or with _____ materials.

   a. design   b. man-made   c. plant   d. solid

2. The _____ an enclosure element is, the more effective it is in traffic control.

   a. wider   b. higher   c. more colorful   d. more solid

3. The type of constructed enclosure selected by a landscaper is usually determined by _____.

   a. the preferences of the client.   c. the money available for the job.
   b. the function it is to perform.   d. the past experience of the designer.

4. Total privacy for two landscapes which are side by side and at the same level requires an enclosure of _____ in height.

   a. 4 feet   b. 5 feet   c. 6 feet   d. 7 feet

5. When the vantage point for a view is highly elevated, the use of _____ helps to block the view.

   a. awnings   b. trees   c. a chain link fence   d. shrubs

6. Weep holes are necessary for _____

   a. wall footings.   c. water pressure release.
   b. structural strength.   d. appearance.

SUGGESTED ACTIVITIES

1. Collect photos illustrating fencing styles to become familiar with what is available for use in landscapes. As sources, use mail order catalogs, garden center and lumber company advertisements, and house and garden magazines. Mount each style separately with its name and several examples of how it can be used in the landscape.

2. Make a similar collection of illustrations of houses. Find examples of as many different styles and construction materials as possible. Match the houses with the most appropriate fencing styles. Avoid mixing styles. Example: Split rail fences look fine with ranch style homes, but are inappropriate with Spanish styling; picket fences go well with Cape Cod styles but not ranch styles.

3. Invite a lumber dealer to talk with the class about wood. Which woods last longest? How can less expensive woods that rot quickly be made to last longer? What prefabricated fencing styles are most popular in your town?

4. Demonstrate the different elevations of enclosure needed to provide total privacy in an area. Have two class members sit opposite each other (perhaps one can sit on a staircase or ladder). Two other class members can then separate them by holding a large blanket or other solid material between them. Measure the height needed to block the students' view of each other. Gradually elevate one student's vantage point, each time measuring the height of enclosure needed for blockage of the view. Alter the placement of the enclosure. Place it closer to the viewer first, and then closer to the person being viewed.

unit 18

SURFACING MATERIALS

OBJECTIVES

After studying this unit, the student will be able to

- select a suitable natural or constructed surface material for a specific area.

- list the advantages and disadvantages of both hard and soft paving.

- determine step dimensions for connecting two different levels of land.

TYPES OF SURFACING MATERIALS

The floor of the outdoor room is formed by the application of surfacing materials. The surfacing is applied only after the soil has been prepared by grading and conditioning. Grading provides for proper water drainage away from buildings and for the movement of water across the soil without harmful erosion. Conditioning of the soil promotes water absorption necessary for the growth of natural surfacings such as grass.

The main reason surfacing is applied is to cover the exposed and disturbed soil surface with a protective material. Without this protection, erosion or compaction of the soil may occur. *Compaction* results when soil particles are pressed tightly together by heavy traffic over the area. Lack of cover can also result in puddling of the soil, which at a severe level can turn a landscape into a mud bath.

There are four basic types of surfacing:

- Paving, for heavy traffic areas

- Turf grass, for areas less subject to use

- Ground covers ⎤
- Flowers ⎦ for areas with no pedestrians

This chart compares different types of surfacing. Numbers which are in parentheses in the chart are explained below it.

Surfacing Type	Installation Cost	Maintenance Cost	Walking Comfort	Use Intensity	Seasonal or Constant Appearance
hard paving	highest	low	lowest	highest	constant
soft paving	moderate	moderate (1)	moderate	moderate (1)	constant
turf grass	lowest	high (2)	high	moderate (6)	constant (7)
ground cover	moderate	moderate (3)	N/A (5)	low	constant to seasonal (8)
flowers	moderate	high (4)	N/A (5)	lowest	seasonal

(1) Some replacement is required each year.
(2) Fertilization, weed control, watering, and mowing are necessary.
(3) Initial cost is high due to the hand weeding which is required. Once established, costs are moderate.
(4) Much hand weeding, watering, and fertilizing are required.
(5) Should not be used in areas with pedestrians.
(6) Use intensity is greatest where people do not continuously follow the same path.
(7) Many grasses are dormant in certain seasons and may change color.
(8) Appearance of these materials depends upon whether plants are evergreen or deciduous.

Paving

Whether or not paving is chosen as a surfacing material depends on how the area in question is to be used. Generally, any area with concentrated foot or vehicular traffic should be paved. Examples are driveways, patios, entries, and areas under outdoor furniture. In these locations, grass quickly wears away, with compaction and mud the result.

There are two subcategories of paving, hard and soft. *Hard paving* includes materials such as brick, stone, poured concrete, tile, paving blocks, and wood planking, figures 18-1, 18-2, and 18-3. The advantages of hard paving are:

Fig. 18-1 Patterned brick ends create an interesting and durable surface.

Fig. 18-2 Brick and concrete are combined for a durable surface with an attractive pattern.

Fig. 18-3 Wood and stone set in mortar give heavy-duty service while complementing the seaside atmosphere.

Greater Durability. Once installed, hard paving provides the longest service at the least cost.

Lower Maintenance Requirements. Usually, only periodic patching is needed to keep the surfacing attractive.

Strength. Where traffic produces great pressure on the surface, hard paving is superior.

Hard paving also has certain disadvantages:

Heat Absorption. After a length of time in the hot sun, hard paving may become hot to the touch. It also releases the heat slowly at night, making patios warmer than desirable on hot summer evenings.

Hazardous When Wet. Not all pavings are slippery, but smooth concrete, tile, and even wood and stone can be hazardous after a rain. For this reason, they should not be used to surface areas near steps or ramps. Likewise, they should not be used for landscapes that are used by ill or handicapped persons.

Glare. Smooth-surfaced hard paving reflects sunlight and can be blinding on bright, clear days. Use of a smooth concrete patio next to a glass door can fill the house with unwanted heat and reflected glare.

Expense. Hard paving is the most costly way to surface the landscape.

Soft paving includes asphalt and a large group of materials known as *loose aggregates*. Some loose aggregates are crushed stone, marble chips, sand, wood chips, bark chips, and tanbark. These pavings are not "soft" in the sense that they would not cause pain if an individual fell on them. While some are actually softer than the hard pavings (such as sand or wood chips), the term *soft paving* indicates that the materials are less durable than hard paving and lack a solid form. They are best used in the landscape where pedestrian or vehicular traffic is not intense, figures 18-4 and 18-5. Soft paving does not perform satisfactorily if it is overused.

Fig. 18-4 Tanbark provides the soft yet durable surface around this playground apparatus.

Fig. 18-5 Crushed stone has been used to surface this traffic circle in a large shopping center. Water can reach plant roots, yet weeds are discouraged.

The advantages of soft paving are:

Lower cost of installation. Neither the cost of materials nor the cost of soil preparation prior to installation is as great as that for hard paving.

Faster Installation. Where the surfacing is needed quickly and/or temporarily, soft paving can be applied more rapidly.

Ease of Application in Oddly Shaped Areas. Soft paving can conform more easily to small or unusually shaped places.

Ease of Replacement. If something is spilled on the surfacing and stains it, it can be easily and inexpensively replaced.

Major disadvantages of the soft paving materials vary with the actual materials. Some of the common disadvantages are:

Greater Maintenance Requirements. Asphalt develops holes. Loose aggregates need to be frequently weeded and raked smooth.

Necessity of Replacing Materials. Loose aggregates are kicked out of place. A small amount is carried away every time someone walks on it; new material must be added annually.

Tendency to Become Sticky (Asphalt) or Dusty (Aggregates). When the weather is hot and dry, soft paving may be tracked into the house, damaging carpeting or floors.

SELECTING THE CORRECT SURFACING

The final decision by a landscape designer or contractor concerning the correct surfacing to use requires an analysis of several factors:

- **Cost of materials and the budget of the client.** The ideal surfacing may be too expensive, requiring the substitution of a second choice which makes use of less expensive material. Landscapers must be up to date on cost trends in building supplies to make accurate cost decisions.

- **Amount of use the surfacing will receive.** If it is in a primary use area (front entry, patio, driveway), hard paving is needed.

If the area receives only secondary use, soft paving or grass may suffice.

- **Aesthetic appearance.** The color and texture of the surfacing should harmonize with the other materials in the landscape.

- **Shape of the area being surfaced.** Some paving materials are fluid and can easily be molded into almost any shape (concrete, asphalt, loose aggregates). Others have definite square, rectangular, hexagonal, or other geometric shapes (bricks, patio blocks, paving stones). These are difficult to cut and are therefore most effective if used in designs that require minimal cutting.

- **Effect upon the building's interior.** Surfacing next to a home or other building should not create reflected glare or heat within the building, nor should it produce mud, dust, or oil that might be tracked into the building.

- **Maintenance required.** The amount of maintenance needed to keep the surfacing attractive should be carefully weighed against the amount of maintenance clients are willing to do themselves or pay to have done.

STEPS AND RAMPS

If the outdoor room has several levels, each level has a separate surfacing. The surfacing may change or remain the same between levels, depending upon the decision of the designer. Regardless, the levels must be connected with steps, ramps, or ramped steps.

The surfacing of the steps or ramps is usually selected to match that used on the levels being connected. However, there may be variations for either aesthetic or practical reasons. For example, it may be desirable to combine materials used in the surfacing and nearby enclosures in the design of the steps. One practical variation might be the necessity of constructing the steps with a more durable material than that used in the surfacing. Another variation might take into account the fact that some materials suitable for surfacing flat areas are slippery as steps or ramps.

A step is constructed of two parts: an elevating portion, or *riser,* and the part on which the foot is placed, or *tread,* figure 18-6.

Outdoor steps are usually not restricted to the same space limitations as indoor steps, figure 18-7 and 18-8. For this reason, treads can be built wider and risers lower to create a stairway that is more comfortable than indoor steps. The design and construction of outdoor steps should be such that users of the land-scape are allowed to maintain the most natural stride pattern possible. This is best accomplished by constructing steps in accordance with a widely accepted formula: the total length of two risers and one tread should equal 26 inches (T + 2R = 26″). Furthermore, the tread should be at least wide enough so that an entire foot can be placed upon it (12-inch minimum). Using this formula, if outdoor steps are built with a 12-inch tread, each riser will be 7 inches high (12″ + (2 x 7″) = 26″), figure 18-9. As the tread dimension increases, the riser dimension decreases:

Fig. 18-6 The parts of a step.

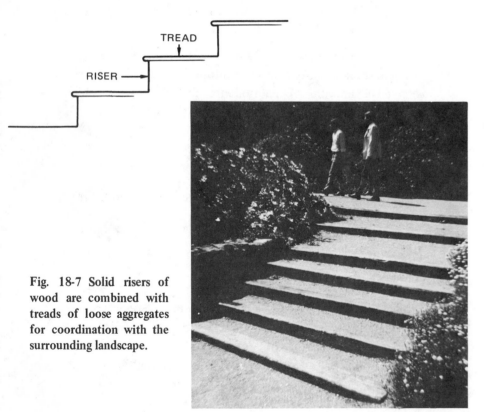

TREAD

RISER

Fig. 18-7 Solid risers of wood are combined with treads of loose aggregates for coordination with the surrounding landscape.

Fig. 18-8 Because of the log risers and earthen treads, these steps blend in comfortably with the woodland setting.

Fig. 18-9 Calculating the dimensions of outdoor steps

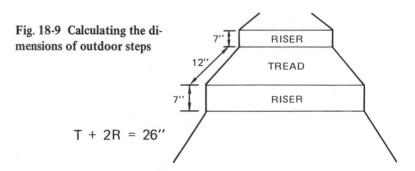

7″ RISER

12″ TREAD

7″ RISER

T + 2R = 26″

riser	tread
7 inches	12 inches
6 inches	14 inches
5 inches	16 inches
4 inches	18 inches

The number of steps required to connect two levels is calculated by dividing the elevation by the riser dimension desired. For example, if two levels are 48 inches apart and a 6-inch riser is desired, eight steps are required (48″ ÷ 6 = 8 steps), figure 18-10. The amount of horizontal space required for the steps is determined by multiplying the tread dimension by the number of steps. In the example, the eight steps, having seven 14-inch treads, require 98 inches (8.16 feet) of space in the landscape for their construction (14″ x 7 treads = 98″ ÷ 12″ = 8.16′). It should be noted that the number of treads is always one less than the number of risers, since the top riser connects with the upper level, not another tread.

Fig. 18-10 Planning space for outdoor steps

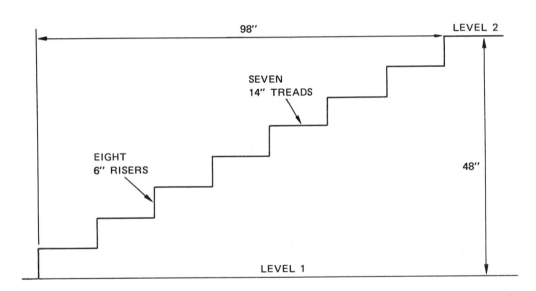

ACHIEVEMENT REVIEW

A. There are four basic types of surfacing: paving (P), turf grass (T), and ground covers and flowers (G/F). Based upon their characteristics as described in this unit, indicate which type is most appropriate in each of the following situations.

1. This surface receives moderate foot traffic. There is no reason for fixed patterns to develop across it.

2. This surface receives heavy vehicular traffic.

3. This surface receives no traffic. It functions as a natural element of the outdoor room.
4. This surfacing must be durable and long lasting. It receives little or no maintenance during the year.
5. This surfacing offers seasonal variation in the landscape.
6. This surface is in a picnic grove where tables are moved to a different spot each week.
7. This surface is in a picnic grove where tables are never moved.
8. This surfacing must tolerate foot traffic while offering a high degree of walking comfort.
9. This surface must be suitable for dancing and shuffleboard.

B. Label the following paving materials as hard or soft.

1. wood rounds
2. crushed stone
3. flagstone
4. wood chips
5. brick
6. poured concrete
7. tanbark
8. asphalt
9. marble chips
10. sand

C. Indicate which paving type, hard or soft, is most suitable in the following situations.

1. The surface is in an eating area and may receive frequent food spills.
2. The surface is next to a heavily used exit from a building.
3. The surface will receive heavy and frequent auto traffic.
4. The surface is located where a dog will play in a service area.
5. The surface must allow rapid drainage of water through it.
6. The surface is to be a patio in a family living area.
7. The surfacing is to be as comfortable to the feet as possible.
8. The surfacing will receive little or no maintenance.
9. The surface must be constructed of the least expensive paving material available.

D. List the six factors which must be considered when selecting surfacing.

E. Using the formula $T + 2R = 26''$, what size riser is best for a step with a 15-inch tread? Show how you find the answer.

F. How many steps having 5-inch risers are necessary to reach an elevation of 5 feet? Show your work.

G. How much horizontal space would the above steps require? Show your work.

H. How high above level B is level A in the following plan view?

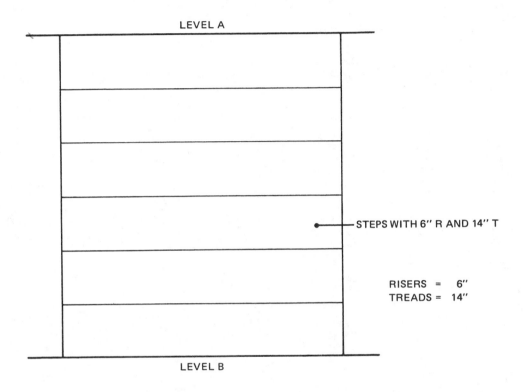

Not applicable

SUGGESTED ACTIVITIES

1. Select an assortment of surfacings near the classroom (turf grass, asphalt, hard paving, and loose aggregates). Rate them according to foot comfort and appearance. Make comparisons in sun and shade, with and without shoes, and early in the day as opposed to late in the day. Are there any observable differences related to the surfacings' color or texture, softness or firmness, and whether the sun is able to reach them or not?

2. Fill waxed milk containers with oven-dried loose aggregates ranging from coarse to fine textures. Punch several holes in the bottom of each container. Compare the speed with which a measured amount of water drains through each aggregate. Measure the amount of water recovered from each. Which aggregate drains most quickly? Which retains the greatest amount of water?

3. Measure the risers and treads of a set of outdoor steps near the school or in a nearby park. Do the measurements follow the formula $T + 2R = 26''$?

section 5

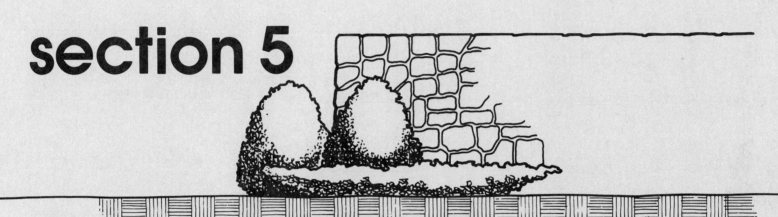

SELECTING ENRICHMENT ITEMS FOR THE LANDSCAPE

unit 19

NATURAL ENRICHMENT ITEMS

OBJECTIVES

After studying this unit, the student will be able to

- explain what a natural enrichment item is, as opposed to other natural elements in the landscape.
- list examples of natural enrichment items commonly available for use in landscaping.

To understand the concept of enrichment, picture again the indoor room. When the basic structure of an indoor room is completed, it has walls, a ceiling, and a floor, but is still lacking in many ways. To make it a usable and personal room, such things as furniture, lighting, pictures, music, and pets must be added. Although these items do not function as walls, ceilings, or floors, they nevertheless play a valuable role.

The outdoor room has a similar need to be made usable and personal. *Enrichment items* are elements of the outdoor room which are not essential to the formation of its walls, ceiling, or floor. *Natural enrichment items* are those elements which have been formed by nature and are either present at the site or moved to the site by the landscaper.

Natural enrichment items may be either *tangible* (touchable) or *intangible* (not touchable).

TANGIBLE ENRICHMENT ITEMS

Stones

Rocks, boulders, and natural outcroppings are valuable additions to any landscape. Of course, the landscape must be sizeable enough to accept the stones without being overpowered by them. The use of stone as a tangible enrichment item is different

Fig. 19-1 This natural stone ledge contributes to the beauty of a lakeside home.

from its use in walls or surfacings. As enrichment items, stones are natural outdoor structures, figures 19-1 and 19-2.

Viewed in a setting where they occur naturally, stone outcroppings can be appreciated for the strong impact they have on the mood of the landscape. The outcroppings of bedrock in New York's Central Park are one famous example. Another example is illustrated in figure 19-3, where the large boulders on a college campus create a spot for quiet conversation.

The Japanese were the first to recognize the enriching quality of stones in the landscape. Japanese gardens are often designed to represent nature in miniature. In this type of design, large stones are used to suggest mountains, animals, or other features. The western world has made great use of the rock garden as a way of combining plants and stones in an enriching manner.

Fig. 19-3 Rock formations add natural enrichment to a college campus.

Fig. 19-2 Large stones provide an enriching backdrop for the rock garden plants growing among them.

Specimen Plants

While the majority of plants serve as wall, ceiling, or surfacing elements in the landscape, some play double roles as tangible enrichment items. Specimen plants may be used as sculpture for their unusual growth habit and for their blossom or foliage color, figure 19-4. When pruned into unusual shapes, the plants become highly styled objects of art, figure 19-5. The pruning of plants into unusual shapes is termed *topiary pruning*. This use of a natural enrichment item creates a high maintenance requirement in the garden. Topiary pruning is done sparingly for this reason.

Water

Whether it occurs as still, quiet pools, figure 19-6, or cascading falls, figure 19-7, water possesses natural qualities that are both tangible and intangible. Water can be touched, but it also can be heard. It moves, reflects light, and sparkles. These are important properties, even though they are not tangible.

Should water exist on a site to be developed, the designer should carefully consider its potential. Water is a strong attraction for people. They like to be near it, to listen to it, to watch it ripple or fall. They fish in it, swim in it, and sail on it. It is the most popular of all recreational and aesthetic enrichments.

Fig. 19-6 A still pond enriches with its own natural beauty and reflection of the surrounding setting.

Fig. 19-4 The unusual silhouette of this spruce gives it strong eye appeal. It becomes an enriching focal feature.

Fig. 19-5 The topiary pruned shrub as an enrichment item requires high maintenance and is used only in very formal settings.

Fig. 19-7 Cascading water provides an enriching quality through its sparkle, sound, and movement.

Fig. 19-8 Water fowl enrich this garden setting.

Animals

Animals give a sense of life to a landscape that no other enrichment item can duplicate, figure 19-8. The songs of birds and the color of their flitting wings, the scurrying of a chipmunk, and the deep croak of a frog all affirm the presence of nature in the outdoor room. Our parks, campuses, and backyards can easily possess some form of wildlife enrichment if a little care is taken to develop a landscape which also acts as a habitat for them. Many plants produce berries which attract birds and other small animals. A hedgerow can supply suitable nesting sites. The addition of feeding stations assures the presence of wildlife throughout the winter as well.

While city and suburban landscapes may require some planning to encourage the presence of wildlife, rural landscapes usually do not. A landscape designer is free to incorporate views of grazing cattle, sheep, horses, or other domestic animals into the farm landscape. In every case, whether urban or rural in setting, the enrichment provided by animal life is natural and desirable.

INTANGIBLE ENRICHMENT ITEMS

Intangible but invaluable are those enrichments which appeal to the senses other than sight. Many plants are noted for their fragrance or for the tasty fruit they produce. What could be a better reason for using them in the landscape? An aromatic (fragrant) plant beneath a bedroom window or near a patio is a sign of thoughtful landscape development.

Also pleasant is the feel of certain enrichment items. White pines, for example, are noted for their soft texture. When wind blows through pine trees, the sound which is produced helps to create a mood that is distinct to the outdoors. Birds and running water, already mentioned as tangible enrichment items, have intangible qualities as well. The landscape designer should attempt to include these intangible natural enrichment features in gardens whenever possible.

ACHIEVEMENT REVIEW

A. Indicate which of the following characteristics apply to enrichment items (E), non-enrichment items (N), or both (B).

1. Their major function is to provide shelter.
2. Their major function is to shape the outdoor room.

3. Their major function is something other than serving as a wall, ceiling, or floor element in the outdoor room.
4. They may be tangible items.
5. They may be tangible or intangible.
6. They could be used as focal points of the design.
7. They could protect people from a rainstorm.
8. They may require periodic maintenance.
9. They are sometimes created by nature.

B. Indicate if the following natural enrichment items are tangible (T) or intangible (I).

a boulder	a waterfall
a lake	the sparkle of a waterfall
the sound of a bird	the sound of a waterfall
a distorted old pine tree	berries on a shrub
wind whistling through trees	the taste of berries

C. Give four examples of natural, tangible enrichment items. Do not use any of those listed in question B.

D. Give four examples of natural, intangible enrichment items. Do not use any of those listed in question B.

SUGGESTED ACTIVITIES

1. List all of the natural enrichment items in your home landscape. Indicate if they are tangible, intangible, or have characteristics of both.

2. Do a similar study of a park near the school. Which landscape has more natural enrichment, the park or the home?

unit 20

MAN-MADE ENRICHMENT ITEMS

OBJECTIVES

After studying this unit, the student will be able to

- list three types of man-made enrichment items available for use in landscape development.
- list five ways in which lighting can be used to create an attractive landscape at night.
- determine the contribution made by an enrichment item.

Enrichment items that are created through a manufacturing process are known as *man-made enrichment items*. Subject to the influence of supply and demand, man-made enrichment items are usually available in a wide range of qualities and prices.

TYPES OF MAN-MADE ENRICHMENT ITEMS

Outdoor Furniture

Just as the indoor room needs furniture to be comfortable and usable, so does the outdoor room. At the residential level, outdoor furnishings may include chairs, lounges, and tables. Park benches, trash containers, playground equipment, and other specialized items are considered furniture in a public landscape.

The landscaper should encourage the client to select outdoor furniture appropriate to the quality of the outdoor room in which it will be placed. For the home landscape, outdoor furniture is sold in many places, ranging from grocery stores to department stores. One familiar type of outdoor chair is made of lightweight aluminum tubing with plastic mesh stripping. It is comparatively inexpensive, moderately comfortable, brightly colored, and lasts two or three seasons. Over a period of several years, more durable furniture might be a better investment, figure 20-1. It may cost more at first, but this type of furniture usually gives greater comfort and satisfaction for a longer period of time.

Fig. 20-1 Metal chairs and tables make attractive yet durable outdoor furniture. In northern areas, this type of furniture must be stored indoors during winter months.

Fig. 20-2 These benches are a permanent landscape item.

In some situations, it is desirable to develop permanent furniture for landscapes. In these cases, it must be durable enough to withstand constant contact with the weather, figure 20-2.

Fig. 20-3 This large sculpture of Prometheus looks down upon skaters at Rockefeller Center in New York City.

Outdoor Art

Many indoor rooms have pictures on walls and sculpture on coffee tables. These enrichment items help to personalize the room and make it different from all others. The outdoor room also benefits from the enrichment of art. In many large cities, the walls of old buildings are looking fresh and new as local artists turn gritty brick into bright multistory murals. Hotels, corporations, and college campuses are making sculpture an important part of their landscapes. Because of this, the public is gaining a greater appreciation of outdoor art in landscapes, figure 20-3.

One problem concerning the addition of good quality, original outdoor sculpture to the landscape is that it is often very expensive. For this reason, the individual situation should be analyzed carefully before sculpture is selected. When the budget allows, the use of original artwork can be highly rewarding, figure 20-4. Good quality reproductions can also be attractive.

Pools and Fountains

There are many ways that water in man-made forms can be used to enrich landscapes. Swimming pools are enjoying greater

Fig. 20-4 This modern sculpture, combined with the falling water, adds original and personal enrichment to the patio.

Fig. 20-5 This large splashing fountain, located in a pedestrian plaza, adds a refreshing coolness to the surroundings.

popularity each year. From large inground installations to small aboveground forms, the pool has caught the interest of many homeowners. The wise designer considers this use of water where space, weather, and the budget permit.

Also adding to the public's appreciation of man-made water features are the large fountains being used in shopping malls and pedestrian plazas, figure 20-5. While some cities of the world have used richly sculptured fountains in their designs for centuries, many smaller towns are beginning to recognize their value to city living.

The residential landscape can also be designed to incorporate the enriching qualities that water offers. Small recirculating fountains, available at a modest price, can bring new life to a patio area. All such fountains are self-contained, electrically powered units that operate by use of a simple wall switch. Also within the price range of most homeowners are small fish ponds such as the one illustrated in figure 20-6.

Student designers today face an exciting challenge: to develop more and better uses of water in the home and commercial land-

Fig. 20-6 A backyard fish pond can bring enjoyment to the smallest family living area.

scape. Water and another enrichment item, night lighting, are among the fastest growing areas of development in the landscaping business.

Night Lighting

When night falls, the enjoyment of the landscape can come to an end if no provision has been made for night lighting. With the addition of imaginative lighting, the night landscape can be made both usable and attractive.

Night lighting is used

- to increase the amount of time a garden may be enjoyed each day.
- to provide safety and security for users of the landscape.
- to create special effects such as colored lighting, silhouette lighting, shadow effects, or patterns against buildings.
- to maintain the same visual relationships between house and garden that exist during the daylight hours.

Until recently, most backyard lighting consisted of floodlights mounted on the roof of the house or garage. The glare of white light was not attractive and often created a prison yard effect. Such floodlights are still excellent security lights, but lighting

Fig. 20-7 Globe lights illuminate while giving the area an aesthetic benefit.

design has advanced to softer, more attractive forms for the outdoor room, figures 20-7 and 20-8.

There are five common methods of night lighting, figure 20-9.

Walk lights offer both safety and decorative effects. They should be used wherever it is necessary to warn pedestrians that the walk is about to change direction or elevation (such as on steps or ramps).

Silhouette lighting outlines plants when placed behind them. The viewer sees a dark plant form against a background of light.

Fig. 20-8 Lighting in the style of an earlier era helps give this small park a restful atmosphere.

Fig. 20-9 Five common outdoor lighting techniques

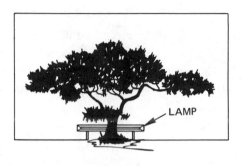

SILHOUETTE LIGHTING. . . THE LAMP IS PLACED BEHIND THE PLANT.

WALK LIGHTS GIVE SPECIAL EFFECTS, MARK CHANGES IN DIRECTION, AND OFFER SAFETY.

SHADOW LIGHTING. . .THE LAMP IS IN FRONT AND THE WALL IS BEHIND THE PLANT.

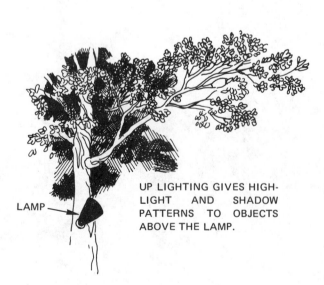

UP LIGHTING GIVES HIGH-LIGHT AND SHADOW PATTERNS TO OBJECTS ABOVE THE LAMP.

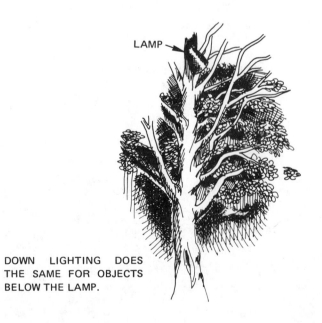

DOWN LIGHTING DOES THE SAME FOR OBJECTS BELOW THE LAMP.

Shadow lighting places the light source in front of the plant and causes a shadow to be cast onto a wall behind the plant.

Down lighting creates patterns of light and leaf shadows on the ground. The light fixture is placed high in a tree and directed downward.

Up lighting is the reverse of the above. The light fixture is placed at the base of the object being illuminated. It is directed upward.

There are two precautions for beginning designers when planning the illumination of landscapes. One is to position the lights so that they do not shine into the eyes of the landscape's users. The other precaution is to be sure that the level of brightness outside the building equals that inside. If the brightness levels are not equal, the separating glass door or window reflects like a mirror. The value of outdoor lighting is then lost to the indoor viewer looking out.

IS THE ENRICHMENT ITEM NECESSARY?

It is possible to design a landscape with too many enrichment items. When this is done, the effect is one of clutter rather than harmonious design. Beginning landscapers may find it difficult to determine the proper number of items to use in a landscape.

A few simple guidelines can be helpful to the new designer. First, be sure the item being considered is truly enriching to the landscape. It must fulfill a role other than the one of walls, ceiling, or surfacing in the outdoor room. Second, most enrichment items attract the viewer's eye strongly and should be used sparingly as focal points in the landscape. Finally, when in doubt concerning the value of an enrichment item, remove it. Then stand back and observe. If nothing appears to be missing in the landscape, it is probably unnecessary. If it leaves a visual hole in the design, the enrichment item probably belongs there.

ACHIEVEMENT REVIEW

A. Indicate whether the following enrichment items are natural (N) or man-made (MM).

1. bird bath
2. sun dial
3. lake
4. statue
5. large boulder
6. chaise lounge
7. picnic table
8. chipmunk
9. outdoor lights
10. wind in pine trees

B. Select the best answer from the choices offered.

1. Where is the light for silhouette lighting located?

   a. behind the plant   b. above the plant   c. in front of the plant

2. Where is the light for shadow lighting located?

   a. behind the plant   b. above the plant   c. in front of the plant

3. What is a good way to test the value of an enrichment item in a landscape?

    a. Sell it and determine its worth.

    b. Remove it temporarily.

    c. Add another one like it.

4. Permanent outdoor furniture should be attractive and _____.

    a. lightweight.   b. weather resistant.   c. upholstered.

SUGGESTED ACTIVITIES

1. Demonstrate different lighting techniques in the classroom. Use a small flashlight for the light source. Use a box to represent the building or wall background. Select a small leafed branch from a shrub to represent the plant. Darken the classroom and try up lighting, down lighting, silhouette lighting, and shadow lighting the plant and building.

2. This activity demonstrates the necessity of designing outdoor lighting so that the level of brightness is close to that of the indoors. Go outside on a bright sunny day and try to see into the school building from the road or sidewalk. The windows, acting as mirrors, will cause the inside of the building to appear dark.

3. Study the sidewalk system around the school building. Where are walk lights needed for safety?

section 6

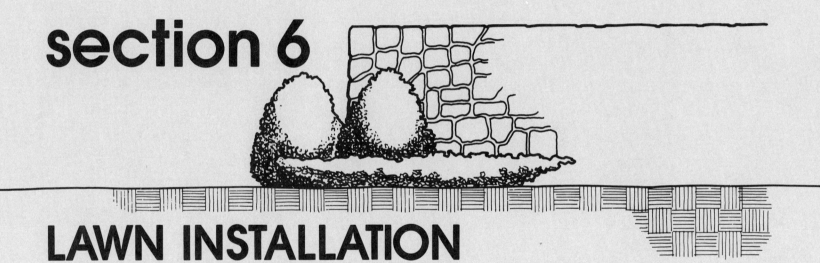

LAWN INSTALLATION

unit 21

SELECTING THE PROPER GRASS

OBJECTIVES

After studying this unit, the student will be able to

- select grass seed for a particular landscape.

- list four characteristics of grass seed which affect its price and quality.

- list the information required by law on grass seed labels.

- explain why a mixture of grasses is often preferable to a single species.

The most commonly used carpet for the outdoor room is the turf grass lawn. Where grass is able to survive, an evenly textured, neatly trimmed green lawn is very attractive. A beautiful lawn, however, is no accident — it requires the same care as any other planting. Quality grass species that are appropriate for the site and proper conditioning of the soil help to assure successful lawn installation.

DIFFERENTIATING AMONG GRASS SEEDS

A trip to a garden center or supermarket in the early spring reveals the nearly endless amount of brands and formulations of grass seed on the market. Names such as "Sure-Grow," "Easy Lawn," "Play Lawn," and "Shade Mix" seem to indicate that there is a grass seed for every situation. There also seems to be a grass seed in almost every price range. The selection of seed is made more confusing by the brightly colored, attractively illustrated packaging of even the cheapest seed. All promise wonderful results in return for the buyer's dollar. It is easy to become confused.

Landscapers must know how grass seed mixtures differ for two reasons:

- To assure that the seed purchased produces the lawn desired in the particular landscape setting

• to be able to explain to their clients why a more expensive seed may be needed for their landscapes

Most grass seeds look alike in their packages. Yet two packages may vary greatly in price. In other cases, the same price may prevail, but the packages vary tremendously in weight. There are several reasons for such variations.

Texture

Texture of the grasses is one reason for variance in price. Certain highly desirable grasses are regarded as fine-leaf forms. *Fine-leaf* grasses have narrow blades and do not become stalky at the base. They result in an evenly textured, attractive lawn which grows at a consistent rate. They are often the end product of years of costly scientific research. Other grasses have a *broad-leaf* form; their blade is much wider and they grow tough and stalky at the base as summer progresses. These grasses create a coarse-textured lawn that is often difficult to mow. They grow unevenly and give a shaggy effect to the lawn. Generally, fine-leaf grasses are considered better lawn grasses and therefore are more expensive than the broad-leaf grasses.

Figure 21-1 categorizes some of the more common lawn grasses.

Size of Seed

The size of the seed is another reason for variation in the quality and quantity of grass seed mixes. Fine-textured grasses have very small seeds. Coarse-textured grass seeds usually are much larger. Thus, a pound of fine-textured grass seed contains considerably more seeds than a pound of coarse-textured grass seed.

Because of the greater number of seeds per pound, a pound of fine-textured grass seed plants a larger area of land. For example, a pound of fine-textured Kentucky bluegrass contains approxi-

Fig. 21-1 Texture chart for grasses

Grass	Permanent	Nonpermanent	Fine Textured	Coarse Textured
Kentucky bluegrass	X		X	
Merion bluegrass	X		X	
Creeping red fescue	X		X	
Chewings fescue	X		X	
Bentgrass	X		X	
Bermuda grass	X		X	
Tall fescue	X			X
Meadow fescue	X			X
Timothy	X			X
Orchard grass	X			X
Redtop		X		X
Annual ryegrass		X		X
Perennial ryegrass	X			X

mately 2,000,000 seeds. That number of seeds plants about 500 square feet of lawn. A pound of coarse-textured tall fescue contains 227,000 seeds; therefore, only 166 square feet can be planted with a pound of this particular seed.

There are other comparisons that further point out the difference in seed sizes. For example, there are as many seeds in 1 pound of bluegrass as there are in 9 pounds of ryegrass; and as many seeds in a pound of bentgrass as there are in 30 pounds of ryegrass.

Noxious Weed Content

Noxious weed content is another measure of the quality of a grass seed mix. *Noxious weeds* are perennial weeds which are especially difficult to control in lawn plantings. They tend to have large seeds which are screened out during the formulation of top quality, fine-textured lawn mixes. As the percentage of coarse-textured grasses in a seed mix increases, the noxious weed content usually increases as well. Noxious weeds are defined by law in most states.

Permanent vs. Nonpermanent Grasses

The percentage of permanent and nonpermanent grasses also affects the quality and price of a grass seed mix. Grasses such as annual ryegrass germinate quickly and provide a fast cover for barren soil. However, they do not reappear the following year. Thus, a grass seed mix which contains a high percentage of annual grasses is of limited value and generally does not cost as much as a mix containing primarily perennial grasses.

WHICH SEED TO BUY?

Although the landscaper may be aware that high quality and low quality seeds are available, purchasing seeds can nevertheless be confusing. This confusion can be avoided by looking beyond the colors and claims on the packages and examining the seed analysis label. The *seed analysis label,* which by law must appear on every package of seeds to be sold, gives a breakdown of the contents of the seed package on which it appears. The analysis label may be on the package itself, or, if the seed is being sold in large quantities, on a label tied to the handle of the storage container.

While legal definitions vary somewhat from state to state, most analysis labels contain the following information:

Purity. The percentage, by weight, of pure grass seed. The label must show the percentage by weight of each seed type in the mixture.

Percent Germination. The percentage of the pure seed which was capable of germination (sprouting) on the date tested. The date of testing is very important and must be shown. If much time has passed since the germination test, the seed is older and less likely to germinate satisfactorily.

Crop Seed. The percentage, by weight, of cash crop seeds in the mixture. These are undesirable species for lawns.

Weeds. The percentage, by weight, of weed seeds in the mixture. A seed qualifies as a weed seed if it has not been counted as a pure seed or a crop seed.

Noxious Weeds. Weeds which are extremely undesirable and difficult to eradicate. The number given is usually the number of seeds per pound or per ounce of weed seeds.

Inert Material. The percentage, by weight, of material in the package which will not grow. In low priced seed mixes, it includes materials such as sand, chaff, or ground corn cobs. Inert material is sometimes added to make the seed package look bigger. At other times, the inert material is already present in the seed and is not removed because the cost involved would raise the price of the seed.

Three sample analyses follow. Study the contents of the blends and determine which mixture would probably cost the most and which the least.

Mixture A

Fine-Textured Grasses	
12.76% red fescue	85% germ.
6.00% Kentucky bluegrass	80% germ.
Coarse Grasses	
53.17% annual ryegrass	95% germ.
25.62% perennial ryegrass	90% germ.
Other Ingredients	
2.06% inert matter	
0.39% weeds — no noxious weeds	

Mixture B

Fine-Textured Grasses	
38.03% red fescue	80% germ.
34.82% Kentucky bluegrass	80% germ.
Coarse Grasses	
19.09% annual ryegrass	85% germ.
Other Ingredients	
7.72% inert matter	
0.34% weeds — no noxious weeds	

Mixture C

Fine-Textured Grasses	
44.30% creeping red fescue	85% germ.
36.00% Merion bluegrass	80% germ.
13.54% Kentucky bluegrass	85% germ.
Coarse Grasses	
None claimed	
Other Ingredients	
5.87% inert matter	
0.29% weeds — no noxious weeds	

It is likely that Mixture C would be the most expensive. It contains the highest percentage of fine-textured grasses, no coarse grasses, and the lowest percentage of weeds. Mixture A would probably cost the least, since it contains a high percentage of coarse-textured grasses, the lowest percentage of fine grasses, and the greatest percentage of weeds. None of the mixtures is very poor in quality, since there are no crop or noxious weed seeds claimed by any.

WHY BUY A MIXTURE?

Grass seed is purchased either as a mixture (such as described previously) or as a single species (such as all Kentucky bluegrass or all chewings fescue). If an evenly textured lawn of consistent height is desired, it might seem logical to plant a single species of grass. In certain cases, this is done. Golf greens, for example, are usually planted with bentgrass because it can be clipped very short and the texture is even and smooth. In the lawns of most homes, businesses, parks, and cemeteries, however, a single species of grass is actually less desirable than a mixture. One reason is that most lawns must grow under several lighting conditions; parts of some lawns, for example, receive full sunlight all day, while other parts receive no direct sunlight. Still other areas, in which trees cover part of the lawn area, receive a combination of sun and shade during the same period.

Certain species of grasses grow best in the sunlight, whereas others grow best in shade. For example, bluegrasses grow well in open, sunny areas, while fescues grow well in partial shade. In most lawns of the temperate zone, a mixture of seed containing both fescue and bluegrass is preferable to a single species of either.

Another reason to avoid growing a lawn from a single species is that grasses, like other plants, are susceptible to insects and diseases. A lawn composed of a single species can be completely destroyed by a single pest. However, a lawn composed of several species of grasses usually cannot be completely ruined by a single pest.

WARM-SEASON AND COOL-SEASON GRASSES

Certain grasses grow best in warm climates; others are more suited to cool regions. As such, grasses are often grouped into two categories: warm-season grasses and cool-season grasses. Figure 21-2 shows the peak growth rate of the two types of grasses. With milder temperatures of 72° to 75° F favoring the growth of cool-season grasses, it is understandable why such grasses are most common in northern regions of the country. A warmer *optimum* (most favorable) temperature of 85° to 90° F is better suited for the warm-season grasses and explains why they are widely used in the southern states. Knowledge of their optimum growing temperatures also explains why northern lawns are often brown and dormant in midsummer when the days are very warm. Likewise, warm-season grasses do not really flourish in early spring and late fall when temperatures fall below the optimum temperature.

Common Warm-Season Grasses	Common Cool-Season Grasses
Bermuda grass	Kentucky bluegrass
zoysia grass	red fescue
centipede grass	colonial bentgrass
carpet grass	ryegrass
St. Augustine grass	
Bahia grass	

Figure 21-3 illustrates the climatic regions of the continental United States which are favorable for the growth of the grasses listed and others. Any seed purchased for planting in a certain climatic region should be composed of the appropriate grasses.

Fig. 21-3 Regions of grass adaptations.

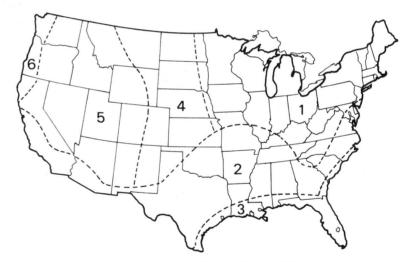

CLIMATIC REGIONS IN WHICH THE
FOLLOWING GRASSES ARE SUITABLE FOR LAWNS:

1. Kentucky bluegrass, red fescue, and colonial bentgrass. Tall fescue, Bermuda, and zoysia grasses in the southern part.
2. Bermuda and zoysia grasses. Centipede, carpet, and St. Augustine grasses in the southern part; tall fescue and Kentucky bluegrass in some northern areas.
3. St. Augustine, Bermuda, zoysia, carpet, and Bahia grasses.
4. Nonirrigated areas: crested wheat, buffalo, and blue grama grasses. Irrigated areas: Kentucky bluegrass and red fescue.
5. Nonirrigated areas: crested wheatgrass. Irrigated areas: Kentucky bluegrass and red fescue.
6. Colonial bentgrass, Kentucky bluegrass, and red fescue.

Fig. 21-2 Relation of temperature to growth rate in cool-season and warm-season grasses.

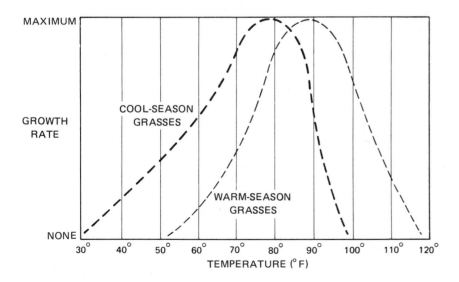

ACHIEVEMENT REVIEW

A.   List four factors that could cause the prices of two 1-pound packages of grass seed to vary greatly.

B.   What could cause a very high quality grass seed purchased in the south to be unsuitable for planting in the north?

C.   Of the three seed mixtures A, B, and C shown in this unit, which mixture is most likely to result in a sparse second-year lawn? Why?

D.   Why is a grass seed mixture usually preferable to a pure, single-species seed?

E.   List and define the important terms found on a grass seed analysis label.

SUGGESTED ACTIVITIES

1.   Grow some grasses. Start flats or coffee cans of pure grass species in the classroom. Compare fine-leaf and broad-leaf types. If possible, also grow samples of warm- and cool-season grasses for comparison.

2.   Obtain several grass seed mixtures from several sources and in as many price ranges as possible. Rank the mixtures on the basis of package appearance, advertised claims, and brand names. Rank the mixtures again, using the seed analysis labels as the measure. How closely do the package claims match the actual facts about the mixture as shown on the labels? How closely does the price ranking follow the quality ranking?

3.   Make a seed count. Weigh 1/4-ounce quantities of a fine-textured grass and a coarse-textured grass. Be as accurate as possible. Count the number of seeds in each measure. Do the fine-textured seeds outnumber the coarse-textured seeds? (Note: Do not use redtop for the coarse-textured grass in this exercise. Its seeds are atypically small for a coarse grass.)

unit 22

LAWN CONSTRUCTION

OBJECTIVES

After studying this unit, the student will be able to

- describe seeding, sodding, and plugging as methods of lawn installation.
- outline the steps required for proper lawn construction.
- explain how to calibrate a spreader.

SELECTING THE METHOD OF LAWN INSTALLATION

Seeding

In the preceding unit, the importance of using good quality seed in lawn installation was emphasized. *Seeding* is unquestionably the most common method of beginning a lawn. It is also the least expensive method. The seed is applied either by hand or with a spreader. The spreader application assures even coverage and is the method used by most professionals, figure 22-1.

Sodding

When a more immediate lawn effect is needed, sodding may be selected as the method of installation. *Sod* is established turf which is moved from one location to another. A sod cutter is used to lift the sod. Then it is rolled up for transport to the site of the new lawn, figure 22-2. Once at the site, it is unrolled onto the conditioned soil bed, figure 22-3. The effect is that of an instant lawn. Sodding is much more costly than seeding, but it produces a more immediate effect. Also, sodding can be used successfully on steep slopes and terraces, where seed might wash away. Thus, sodding is sometimes necessary despite its cost.

Plugging

Plugging is a common method of installing lawns in the southern sections of the United States. Certain grasses, such as

Fig. 22-1 The use of a spreader assures even distribution of grass seed.

Fig. 22-2 A sod cutter removes the growing turf along with a thin layer of soil. Rolled sod (at left) is ready to be transported to a new location.

Bermuda, St. Augustine, and zoysia, are not usually reproduced from seed. Instead, they are usually placed into the new lawn as plugs of live, growing grass. Since the growing season in the southern regions is longer than elsewhere, the plugs have time to develop into a full lawn. Plugging is a time-consuming means of installing a lawn, which is its major limitation.

PROPER LAWN CONSTRUCTION

If the lawn is to be of the best quality, it must be given every possible chance for success. Proper construction of the lawn is vital. Six steps should be followed by the landscaper to assure a successful beginning for the lawn.

- Plant at the proper time of year.
- Provide the proper drainage and grading.
- Condition the soil properly.
- Apply fresh, good quality seed, sod, or plugs.
- Provide adequate moisture to promote rapid establishment of the lawn.
- Mow the new lawn to its correct height.

Fig. 22-3 The sod is unrolled at the new location and laid in place. The effect is one of instant lawn.

Time of Planting

Lawns in southern sections of the country require warm-season grasses. Such grasses grow best in day temperatures of approximately 85° F. It is most effective to plant them in the spring, just prior to the summer season. In this way, they have the opportunity to become well established before becoming dormant in the winter.

Cooler northern regions require cool-season grasses to yield the most attractive lawns. Bluegrasses and fescues germinate best when the night temperatures are in the range of 43° to 70° F. These lawns thrive in locations where days are cool and nights are warm. The best planting time for these grasses is early fall or very early spring, prior to the ideal cool season in which they flourish. If cool-season grasses are planted too close to the intensely hot or cold days of summer and winter, they will die or become dormant before becoming well established.

Grading and Draining the New Lawn

Each time the rain falls or a sprinkler is turned on, water moves into the soil and across its surface. *Grading* (leveling land so that it slopes) directs the movement of the surface water. *Drainage* allows the water to move slowly down into the soil to prevent erosion or puddling.

Even lawns that seem flat must slope enough to move water off the surface and away from nearby buildings. If a slight slope does not exist naturally, it may be necessary to construct one. A *fall* (grade) of between 6 inches and 1 foot over a distance of 100 feet is required for flat land to drain properly. Failure to grade lawns away from buildings can result in flooded cellars and basements.

Drainage of water into and through the soil is important. Without a supply of water to their roots, neither grasses nor any other plants can live. Without water drainage past their roots, turf grasses and other plants can be drowned. Depending upon the soil in the particular lawn area involved, good drainage may require nothing more than mixing sand with the existing soil to

Fig. 22-4 Drainage tile installation. (Drawing not to scale.)

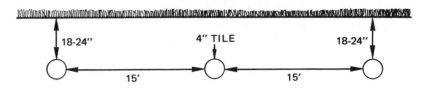

Fig. 22-5 The flow of water through tile drainage system. (Drawing not to scale.)

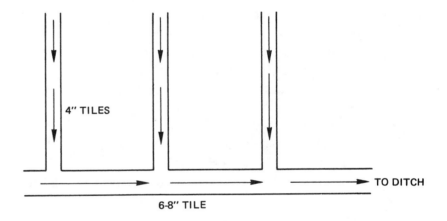

allow proper water penetration. In cases where the soil is heavy with clay, a system of drainage tile may be necessary.

If drainage tile is needed, it should be installed after the lawn's grade has been established, but before the surface soil has been conditioned. Regular 4-inch agricultural tile is normally used, placed 18 to 24 inches beneath the surface. Tile lines are spaced approximately 15 feet apart, figure 22-4. Each of the lateral tiles runs into a larger main drainage tile, usually 6 to 8 inches in diameter. This, in turn, empties into a nearby ditch or storm sewer, figure 22-5.

Where the soil is naturally sandy, no special consideration for drainage may be necessary.

Conditioning the Soil

Proper soil preparation requires an understanding of soil texture and soil pH.

Soil texture is the result of differing amounts of sand, silt, and clay in the composition of the soil. A soil which has nearly equal amounts of sand, silt, and clay in it is called a *loam* soil. Loam soils are excellent for planting. Soil textures such as *sandy loam, clay loam,* and *silty clay loam* are named for the ingredient or ingredients which make up more than one-third of the composition of the soil. For example, the composition of sandy loam is more than one-third sand. Silt and clay each make up more than one-third of the composition of silty clay loam, and hence, it is less than one-third sand. In conditioning soil for lawn construction, clay, sand, silt, or humus (organic material) may be added to bring the existing soil closer to a medium-loam texture.

Soil pH is a measure of the acidity or alkalinity of soil. A pH measurement of less than 7.0 indicates increasing soil acidity. As the pH increases beyond 7.0, the soil becomes more alkaline or basic. Most turf grasses grow best in soil with a neutral pH (expressed as 7.0) to slightly acidic pH (6.5).

The measurement of soil pH is obtained from a soil test. Soil tests are usually available through county Cooperative Extension Services. Also, pH test kits can be purchased at a reasonable cost, allowing landscapers to make their own determination of pH much more quickly.

If the pH of soil is too acidic, it is usually possible to raise the pH by adding dolomitic limestone. The limestone should be applied in the spring or fall. The amount applied depends upon the texture of the soil and how far the natural pH is from 6.5 to 7.0.

Where it is necessary to lower the pH of the soil to attain the desired level of 6.5 to 7.0, landscapers commonly use sulfur, aluminum sulfate, or iron sulfate.

The attainment of a suitable texture and the proper pH are very important in the conditioning of lawn soil. Equally important is the removal of stones from the surface layer of the soil; the loosening of the soil to a depth of 5 or 6 inches; and the incorporation of organic matter into the soil.

Natural Soil	Soil Texture		
	Sandy	Loam	Clay or Silt
pH 4.0	90 lb	172 lb	217 lb
pH 4.5	82 lb	157 lb	202 lb
pH 5.0	67 lb	127 lb	150 lb
pH 5.5	52 lb	97 lb	120 lb
pH 6.0	20 lb	35 lb	60 lb
pH 6.5-7.0	None needed	None needed	None needed

Amount of Dolomitic Limestone Applied Per 1,000 Square Feet of Lawn

Stones may be removed by hand, by rake, or by machine. If the lawn is to be smoothly surfaced, even the smallest surface stones must be discarded.

Decaying organic matter creates *humus,* a valuable ingredient of soil. Humus aids the soil in moisture retention. It also helps air reach the soil. Organic matter can be added to the soil during its conditioning with materials such as peat moss, well-rotted manure, compost, or digested sewage sludge. The landscaper may choose the material which is easily available and relatively low in cost.

All necessary soil additives (pH adjusters, organic matter, sand, and fertilizers) can be worked into the soil at the same time. This is done most effectively with a garden tiller, which also loosens the soil surface and breaks the soil into small particles, figure 22-6. Once the soil has been properly conditioned, it is ready to plant.

Planting the Lawn

If planting is done with plugs, the method of application is similar to that of ground covers. If done with sod, the turf is rolled out and fitted like an interlocking puzzle. Going over the lawn with a light roller after sodding assures contact of the roots with the soil.

If planting is done with seed, spreaders give the most even distribution. The seed is often mixed with a carrier material such as sand or topsoil to assure even spreading. The mixed seed is divided into two equal parts. One part is sown across the lawn in one direction. The other is then sown across the lawn perpendicular to the first sowing, figure 22-7.

The addition of a light mulch of weed-free straw over the seed helps retain moisture. It also helps to prevent washing away of the seed during watering or rainfall. On a slope, it is wise to apply erosion netting over the mulched seed to further reduce the danger of the seed washing away, figure 22-8. One recent innovation, termed hydroseeding, applies seed, water, fertilizer, and mulch simultaneously. It is commonly used for the seeding of large areas, as along highways.

The Importance of Watering

Water is essential to the growth of all plants. As long as grass is dormant in the seed, it needs no water. However, once planted and watered, the seed swells and germinates. At that point, an uninterrupted water supply is very important. The soil surface must not be allowed to dry until the grass is about 2 inches tall. Watering several times a day, every day over a period of a month may be necessary.

Caution should be taken to keep the new seedlings moist without saturating the soil. Too much moisture can encourage disease development. The use of a lawn sprinkler is a much better

Fig. 22-6 A large garden tiller turns the soil while working soil additives into it. For large lawn areas, the tiller is a necessity.

Fig. 22-7 Spreader application. Half the material is applied at a 90° angle over the other half.

Fig. 22-8 Erosion netting is used here to prevent the grass seed from washing away until it can become established. Steep slopes such as this one are usually difficult to seed.

method than simply turning a garden hose onto the new grass. With a sprinkler, the water can be applied slowly and evenly.

The First Mowing

The first mowing of a new lawn is an important one. The objective is to encourage horizontal branching of the new grass plant as quickly as possible. This in turn creates a thicker lawn faster. The first mowing should take place when the new grass has reached a height of 2 1/2 to 3 inches. It should be cut back to a height of 1/2 to 3/4 inch. Thereafter, different species require differing mowing heights to properly maintain the lawn. For the first mowing, it is good practice to catch the grass clippings. This prevents the buildup of *thatch* (dead, undecomposed grass) on the soil's surface.

CALIBRATING A SPREADER

To *calibrate* a spreader is to adjust it so that it dispenses material (seed, fertilizer, or granular weed killer) at the rate desired. For materials that are applied often by the landscaper, the calibration need be done only once, with the proper setting noted on the control for future reference. However, when different materials are used, even though the rate of application may be the same, a different calibration is usually necessary.

The object of calibration is to measure the amount of material applied to an area of 100 square feet. A paved area such as a driveway or parking lot is an excellent calibration site. Afterwards, the seed or other material can be swept up easily for future use. Covering the area with plastic is also helpful in recollecting the material.

The spreader should be filled with exactly 5 pounds of the material being applied. Selection of a spreader setting near the center of the range is a good point at which to begin.

The material is applied by walking at a normal pace in a straight line. The spreader is shut off while it is being turned around. Each strip should slightly overlap the previous one. When an area of 100 square feet has been covered once, the spreader is shut off. The material remaining in the spreader is then emptied out and weighed. By subtracting the new weight from the original weight, the quantity of material applied per 100 square feet is determined. The spreader can then be adjusted to increase or reduce the rate of application.

ACHIEVEMENT REVIEW

A. Briefly answer each of the following.

How does the cost of sodding compare to that of seeding?

Which method has a more immediate effect, seeding or sodding?

Which lawns are commonly started by plugging?

Which method of lawn installation and establishment requires the most time?

At what time of year should warm-season grasses be planted?

At what time of year should cool-season grasses be planted?

At what time of year should bluegrasses and fescues be planted? Why?

Why is it important that soil drain properly?

9. What size of agricultural drainage tile is recommended for lawn use, and how is it spaced?

10. Define soil texture.

11. What type of soil is considered ideal for planting?

12. Explain soil pH.

13. What is a neutral pH level?

14. If soil pH is raised, does the soil become more acidic or more alkaline?

15. If a sandy soil has a pH of 5.0, how many pounds of dolomitic limestone per 1,000 square feet are needed to raise the pH level to that required for a lawn?

16. If the soil mentioned in question 15 covers a lawn area of 3,000 square feet, how much limestone should the landscaper purchase?

17. What are the water requirements of a new lawn?

18. At what height should a lawn mower be set for the first mowing of a new lawn?

19. What is meant by the calibration of a spreader?

20. How is a spreader calibrated?

SUGGESTED ACTIVITIES

1. Invite a Cooperative Extension Service agent to visit the class for a discussion of soil testing. Ask the agent to demonstrate how a soil sample is collected and to explain how landscapers in the state can arrange to have soil tested.

2. Obtain several inexpensive pH testing kits. Bring in soil samples from gardens for testing.

3. Construct a lawn. If materials for proper lawn construction are available at the school, install a new lawn there. If budget restrictions prevent this, volunteer as a class to be a work force for a nearby park, institution, or property owner in return for equipment and materials for the project.

4. Visit a sod farm if one is located in the area.

5. Borrow several spreaders for calibration. (Families of students might be one source.) If there is no budget for seed, substitute sand for demonstration purposes.

section 7

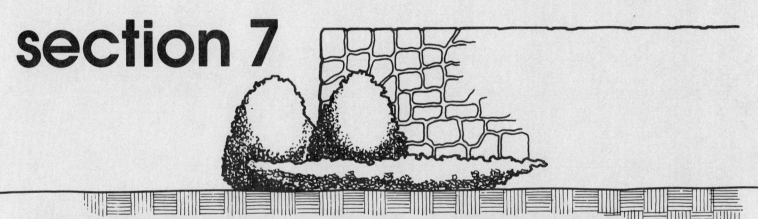

DEVELOPING COST ESTIMATES

unit 23

PRICING THE PROPOSED DESIGN

OBJECTIVES

After studying this unit, the student will be able to

- list the features of a design cost estimate.
- prepare a design cost estimate.

A *cost estimate* is an itemized breakdown of the expenses involved in developing a landscape. It is important and useful to both the landscaper and the client. Naturally, clients want to know the total cost of the proposed landscape for their property. By having an accurate cost analysis in advance, the client can budget for the costs, request revision of certain features, or if necessary, reject the project entirely as too expensive.

The landscape architect, designer, or contractor needs the cost estimate to assure that the client's budget is not exceeded. If it becomes necessary to cut back certain costs, the estimate allows each expense to be examined separately.

FEATURES OF A DESIGN COST ESTIMATE

In a cost estimate, every physical item required for the landscape and its installation cost must be included. Also, every service required in the development of the landscape along with its cost must be listed in the estimate. Specifically, a design cost estimate includes the following:

- Cost of the plant materials
- Cost of installing the plant materials
- Cost of the construction materials
- Cost of installing the construction materials
- Cost of turf grass installation (including materials and labor)
- Fee for landscape designing

To be of greatest value, the cost estimate should be developed at the same time as the design. This assures that the designer is aware of the cost of materials as they are specified for the design. An accurate estimate also requires knowledge of local labor costs.

Staying up to date on material and labor costs is a difficult task. Cost information is easier to obtain for the designer who is also a landscape contractor. Experienced designers maintain an extensive collection of product catalogues and price lists along with a list of sources of supply for construction materials.

PREPARING THE ESTIMATE

The cost estimate should be complete and concise. It should also be understandable and easy to read. All of the items of the estimate should be labeled, with their costs in aligned columns. Subtotals for each category should be set off from the rest of the information. The following are two partial estimates, including an explanation of each item listed.

COST OF PLANT MATERIALS

Number (1)	Species (2)	Description (3)	Cost Per Plant (4)	Total (5)
2	*Acer platanoides*	5 ft./B & B	$20.50	$41.00
1	*Betula pendula*	5 ft./B & B	12.50	12.50
4	*Cornus florida*	3-4 ft./B & B	10.00	40.00
15	*Euonymus alatus*	2 ft./BR	6.00	90.00
200	*Vinca minor*	Rooted cutting	0.40	80.00
			Total Cost of Plant Materials: (6)	$263.50

Note:
(1) *Number* is the total of each separate species used in the design.
(2) *Species* denotes the actual plant being tallied. While either botanical or common names can be used, one should be used consistently.
(3) *Description* gives an explanation of what the supplier is to provide. It also provides a check for the landscaper and a safeguard for the client's investment. The size (such as *5 ft.* or *4-inch caliper*) refers to the plant's size at the time of purchase, not its ultimate mature size. *B & B* stands for balled and burlapped; *BR* stands for bare rooted. They describe the form of root system with which the plant is purchased. Containerized (*CONT*) or other forms might also be specified.
(4) *Cost per plant* is the cost of one plant fitting the description given.
(5) *Total* is the cost of one species of plant multiplied by the number of that species to be purchased.
(6) *Total cost of plant materials* is the sum total of all the materials described in the category. It is set off to the right and underlined twice.

COST OF CONSTRUCTION MATERIALS

Amount (1)	Description (2)	Unit Cost (3)	Total (4)
40 ft.	Redwood stained, post and rail fencing, 3 ft. high	$ 10.00/10 ft. section	$ 40.00
200	Poured concrete patio blocks	1.25/block	250.00
1	Recirculating fountain, pump and basin	350.00	350.00
400 sq. ft.	Poured concrete over 4-in. coarse gravel base	1.00/sq. ft.	400.00

Total Cost of Construction Materials: (5) $1,040.00

Note: (1)   *Amount* is the quantity and unit of measurement (where applicable) of materials being ordered.

(2)   *Description* is the specification of the actual materials to be ordered. It should be detailed enough to prevent error or confusion.

(3)   *Unit cost* is the price of the smallest amount or form of the construction item available. It might be a dozen, a square foot, or a single piece.

(4)   *Total* is the unit cost multiplied by the number of units of each item needed.

(5)   *Total cost of construction materials* is the sum total of all the materials described in the category. It is set off to the right and underlined twice.

The completed cost estimate includes the client's name and address, the name of the designer, and the date of preparation. It is best for the cost estimate to be typewritten and double spaced. The landscape designer should not rely on color coding to distinguish any items, since colors do not show up on carbons or photocopies. The cost estimate always accompanies the presentation of the design to the client.

The following is a completed cost estimate.

COST ESTIMATE FOR THE DESIGN AND DEVELOPMENT OF THE PROPERTY OF MR. AND MRS. JOHN DOE 1234 N. MAIN STREET, CHICAGO, ILLINOIS

I. PLANT MATERIALS TO BE INSTALLED

Number	Species	Description	Cost Per Plant	Total
2	*Acer platanoides*	5 ft/B & B	$20.50	$41.00
1	*Betula pendula*	5 ft/B & B	12.50	12.50
4	*Cornus florida*	3-4 ft/B & B	10.00	40.00
15	*Euonymus alatus*	2 ft/BR	6.00	90.00
200	*Vinca minor*	Rooted cutting	0.40	80.00

Total Cost of Plant Materials: $ 263.50

II. COST OF INSTALLATION OF PLANT MATERIALS

50% of the net cost ($263.50) $ 131.75

III. CONSTRUCTION MATERIALS TO BE INSTALLED

Amount	Description	Unit Cost	Total
40 ft.	Redwood stained, post and rail fencing, 3 ft. high	$ 10.00/10 ft. section	$ 40.00
200	Poured concrete patio blocks	1.25/block	250.00
1	Recirculating fountain, pump and basin	350.00	350.00
400 sq. ft.	Poured concrete over 4-in. coarse gravel base	1.00/sq. ft.	400.00

Total Cost of Construction Materials: $1,040.00

IV. COST OF INSTALLATION OF CONSTRUCTION MATERIALS

40 work hours estimated @ $12.00/hr. including equipment. $ 480.00

V. INSTALLATION OF TURF GRASS (Materials and Labor)

5,000 sq. ft., seeded, @ 11¢/sq. ft. $ 550.00

VI. TOTAL COST OF ALL MATERIALS AND INSTALLATION (Note 1)   *$2,465.25

VII. FEE FOR LANDSCAPE DESIGNING @ 10% of Item VI $ 246.53

VIII. TOTAL COST FOR COMPLETE LANDSCAPE DEVELOPMENT $2,711.78
(Note 2)

Prepared by Joseph Smith   July 11, 1983

Note: (1) The total cost figure is obtained by adding all of the double underlined figures in the categories. The asterisk distinguishes it from other figures.

(2) The final figure should be the most distinctive on the page. It should be set off from the other figures in some way, or at least be the bottom figure on the page, with no others below it to create confusion.

PRACTICE EXERCISE

Prepare a complete cost estimate for the property of Mr. and Mrs. Dwayne Johnson, 1238 N. Grand Street, Brownsville, California. Use the data below. Follow the format outlined in this unit.

PLANT MATERIALS TO BE PURCHASED

4 Chinese redbud @ $12.50/5 ft. tall/B & B
20 golden forsythia @ $5.00/2 ½ ft. tall/BR
2 sugar maple @ $7.50/7 ft. tall/BR
12 Carolina rhododendron @ $15.00/3 ft. tall/B & B
10 Exbury azaleas @ $7.50/3-4 ft. tall/CONT
20 Van Houtte spirea @ $5.00/2 ½ ft. tall/BR
300 Japanese pachysandra @ 40¢/rooted cutting
1 Babylon weeping willow @ $6.00/6 ft. tall/BR
12 hybrid tea rose @ $5.50/1 ½-2 ft. tall/CONT
2 red oak @ $15.00/5-6 ft. tall/B & B

CONSTRUCTION MATERIALS TO BE PURCHASED

4,000 brown brick @ 10¢ per brick
800 sq. ft. poured concrete @ 50¢ per sq. ft.
40 ft. rustic rail fence/3 ft. tall @ $1.25 per ft.
5 walk lights @ $50.00 per light
200 sq. ft. red outdoor carpet @ $1.00 per sq. ft.
1 ½ tons wood chips @ $60.00 per ton untreated

AMOUNT AND METHOD OF LAWN INSTALLATION

6,000 sq. ft., sodded @ 18¢ per sq. ft.

LABOR AND SERVICE CHARGES

Cost of installing plant materials: 50% of the total net value of plant materials
Cost of installing construction materials: two and one-half times the total net value of construction materials
Fee for landscape designing: 15% of the total cost of materials and installation

ACHIEVEMENT REVIEW

A. What is a cost estimate?

B. Who benefits from a cost estimate? How?

C. What are the six features of the design cost estimate?

D. Define the following terms as applied to cost estimates:

   1. B & B   2. cost per plant   3. unit cost

E. Unscramble the data below and arrange it in the proper form for a cost estimate. Use the examples given in the unit for guidance.

   The following materials are needed for a landscape's development:

   1. Eighteen *Cornus stolonifera* are purchased. Each costs $5.00 and is 1 1/2 feet tall. All are purchased bare rooted.
   2. Five *Acer saccharum* are purchased. They are 7 feet tall and balled and burlapped. The total cost is $50.00.
   3. Seven slender deutzia *(Deutzia gracilis)* are used. All are containerized and cost $3.50 each.
   4. The ground cover, *Hedera helix,* costs 40 cents per rooted cutting. The total cost is $50.00.

F. Unscramble the data below and arrange it in the proper form for a cost estimate. Use the examples given in the unit for guidance. Show all necessary arithmetic for totals.

   The following construction materials are needed for a landscape's development.

   1. One hundred feet of chain link fence, 8 feet high are used. The cost is $20.00 per 10 feet.
   2. Crushed stone is purchased in a 1 1/2 ton quantity. The stone costs $80.00 per ton.
   3. One thousand red bricks are used. Each brick costs 10 cents.
   4. Two hundred dollars are spent for flagstone to cover an area 100 square feet in size.

unit 24

PRICING LANDSCAPE MAINTENANCE

OBJECTIVES

After studying this unit, the student will be able to

- list the features of a landscape maintenance cost estimate.
- prepare a maintenance cost estimate.

THE NEED FOR COST ESTIMATES IN LANDSCAPE MAINTENANCE

An important part of professional landscape maintenance is knowing the cost of various jobs and, therefore, how much clients should be charged. The failure to figure costs accurately can result in loss of profits for the landscaper or overcharging the customer.

If the landscape grounds keeper is responsible (either by contract or by full-time employment) for projecting budget needs for the next year, cost estimates are valuable aids. The independent landscaper uses cost estimation as a way of comparing the time and profit involved in different jobs. In this way, certain types of maintenance may be emphasized as more profitable. Other tasks may be dropped because they are too time consuming for the limited profit involved.

FEATURES OF THE MAINTENANCE COST ESTIMATE

An accurate cost estimate for landscape maintenance services requires highly detailed record keeping. The landscaper must know the type and quantity of materials necessary for each job. Only then can its cost be accurately measured. When materials are listed, they must be related to individual jobs. For example, a listing might read *cost per thousand square feet of coverage*, but it would not read *cost per bag* or *cost per truckload*.

Similarly, the labor time for each job must be converted to minutes per thousand square feet. Simply measuring a worker by the amount of work accomplished during an hour or day is not adequate for cost estimation.

Specifically, a cost estimate for a maintenance job includes

- a listing of all services to be performed.

- the total square footage area involved for each service.

- the number of times each service is to be performed during the year.

- the number of minutes required to complete each service over an area of 1,000 square feet.

- the total yearly (annual) time required for each service (recorded in minutes).

- the cost of all materials required for each service. (Costs are converted to amount per 1,000 square feet.)

- the labor costs for each service performed.

CALCULATIONS FOR COST ESTIMATES

To apply all of the data properly requires practice. Study the following examples and their explanations.

Example 1

Problem: To calculate the cost of mowing 10,000 square feet of lawn with an 18-inch power mower 30 times each year.

Necessary Information: It takes 5 minutes to mow 1,000 square feet. The laborer receives $4.00 per hour. There are no material costs.

Solution:

Maintenance Operation	Sq. Footage Area Involved	Number of Times Performed Annually	Minutes Per 1,000 Sq. Ft.	Total Annual Time in Minutes (1)	Material Cost Per 1,000 Sq. Ft.	Total Material Cost	Wage Rate Per Hour	Total Labor Cost (2)	Total Cost of Maintenance Operation Per Year
Lawn mowing with 18-inch power mower	10,000 sq. ft.	30	5	1,500	none	none	$4.00	$100.00	$100.00

Note: (1) To obtain the total annual time in minutes:
   a. divide the square footage of area involved by 1,000 [10,000 square feet ÷ 1,000 = 10]
   b. multiply by minutes per 1,000 square feet [10 x 5 = 50 minutes]
   c. multiply by number of times performed annually [50 minutes x 30 = 1,500 minutes]
   d. enter answer under total annual time in minutes.
 (2) To obtain the total labor cost:
   a. divide the total annual time in minutes by 60 minutes [1,500 minutes ÷ 60 minutes = 25 hours]
   b. multiply by the wage rate per hour [25 hours x $4.00 = $100.00]
   c. enter answer under total labor cost.

Example 2

Problem: To calculate the cost of mulching 2,000 square feet of planting beds with wood chips.

Necessary Information: The task is done once each year. It requires 30 minutes to mulch 1,000 square feet. The laborer receives $4.00 per hour. The wood chips cost $20.00 per 1,000 square feet of coverage.

Solution:

Maintenance Operation	Sq. Footage Area Involved	Number of Times Performed Annually	Minutes Per 1,000 Sq. Ft.	Total Annual Time in Minutes (1)	Material Cost Per 1,000 Sq. Ft.	Total Material Cost (2)	Wage Rate Per Hour	Total Labor Cost (3)	Total Cost of Maintenance Operation Per Year (4)
Mulching plantings with wood chips	2,000 sq. ft.	1	30	60	$20.00	$40.00	$4.00	$4.00	$44.00

Note:
(1) To obtain the total annual time in minutes:
   a. divide the square footage of area involved by 1,000 [2,000 square feet ÷ 1,000 = 2]
   b. multiply by minutes per 1,000 square feet [2 x 30 = 60 minutes]
   c. multiply by number of times performed annually [60 minutes x 1 = 60 minutes]
   d. enter answer under total annual time in minutes.

(2) To obtain the total material cost:
   a. divide the square footage of area involved by 1,000 [2,000 square feet ÷ 1,000 = 2]
   b. multiply by the material cost per 1,000 square feet [2 x $20.00 = $40.00]
   c. multiply by the number of times performed annually [$40.00 x 1 = $40.00]
   d. enter answer under total material cost.

(3) To obtain the total labor cost:
   a. divide the total annual time in minutes by 60 minutes [60 minutes ÷ 60 minutes = 1 hour]
   b. multiply by the wage rate per hour [1 hour x $4.00 = $4.00]
   c. enter answer under total labor cost.

(4) To obtain the total cost of maintenance operation per year:
   a. add total material cost and total labor cost [$40.00 + $4.00 = $44.00]
   b. enter answer under total cost of maintenance operation per year.

THE COMPLETED COST ESTIMATE

A full cost estimate for maintenance is simply an enlargement of the previous examples. For convenience, all of the maintenance operations that deal with the same area of the landscape are grouped together in the estimate. Study the following example and note the calculation of all figures.

Data

I. A landscape requires the following maintenance tasks and equipment:

a. 24,500 square feet of lawn cut 30 times each year with a power riding mower

b. 500 square feet of lawn cut 30 times each year with an 18-inch power hand mower

c. all lawn areas fertilized twice each year

d. 5,000 square feet of shrub plantings fertilized once each year

e. shrubs pruned once each year

f. 400 square feet of flower beds requiring soil conditioning once each spring

g. flowers planted once each year

h. flowers hand weeded 10 times each year

i. flower beds cleaned and prepared for winter once each autumn

II. Calculated time requirements for the maintenance tasks:

a. a power riding mower cutting 1,000 square feet of lawn in 1 minute

b. an 18-inch power mower cutting 1,000 square feet of lawn in 5 minutes

c. lawn fertilization requiring 3 minutes per 1,000 square feet for spreading

d. shrub fertilization requiring 5 minutes per 1,000 square feet

e. pruning time for shrubs averaging 60 minutes per 1,000 square feet

f. soil conditioning for flower beds requiring approximately 200 minutes per 1,000 square feet

g. flower planting requiring 600 minutes per 1,000 square feet

h. weeding of flowers requiring 60 minutes per 1,000 square feet

i. cleanup of flower beds in the autumn requiring 400 minutes per 1,000 square feet

III. All laborers receive wages of $3.50 per hour.

IV. Material costs:

a. lawn fertilizer at $4.00 per 1,000 square feet

b. shrub fertilizer at $2.00 per 1,000 square feet

c. conditioning materials for flower beds at $10.00 per 1,000 square feet

d. flowers for planting averaging $75.00 per 1,000 square feet

Maintenance Operation*	Sq. Footage Area Involved*	Number of Times Performed Annually*	Minutes Per 1,000 Sq. Ft.*	Total Annual Time In Minutes**	Material Cost Per 1,000 Sq. Ft.*	Total Material Cost**	Wage Rate Per Hour*	Total Labor Cost**	Total Cost of Maintenance Operation Per Year**
Lawn									
Mowing - rider	24,500 sq. ft.	30	1	735	None		$3.50	$42.88	$ 42.88
Mowing - 18" power	500 sq. ft.	30	5	75	None		$3.50	$ 4.38	$ 4.38
Fertilization	25,000 sq. ft.	2	3	150	$ 4.00	$200.00	$3.50	$ 8.75	$208.75
Shrubs									
Fertilization	5,000 sq. ft.	1	5	25	$ 2.00	$ 10.00	$3.50	$ 1.44	$ 11.44
Pruning	5,000 sq. ft.	1	60	300	None		$3.50	$17.50	$ 17.50
Flowers									
Soil conditioning	400 sq. ft.	1	200	80	$10.00	$ 4.00	$3.50	$ 4.66	$ 8.66
Planting	400 sq. ft.	1	600	240	$75.00	$ 30.00	$3.50	$14.00	$ 44.00
Hand weeding	400 sq. ft.	10	60	240	None		$3.50	$14.00	$ 14.00
Autumn cleanup	400 sq. ft.	1	400	160	None		$3.50	$ 9.31	$ 9.31
									$360.92

Notes: *All entries in this column came directly from the data given.

**All entries in this column were calculated using methods described in the earlier examples. Students should practice the calculations to insure their understanding of the methods.

V. Overhead and Profit

Before the cost estimate is quoted to a potential client, additional charges for *overhead* and *profit* must be added. Overhead costs include all of the operational expenses incurred by the business unrelated to any particular job (utilities, office salaries, stationery, etc.). It may be estimated as a percentage of the total cost of the maintenance operation. Profit may also be calculated as a percentage of the total cost of the maintenance operation. In this example, overhead costs of 10% ($36.09) and profit allowance of 25% ($90.23) could be added to the $360.92 cost of maintenance. The price quotation to the client would then be $487.24.

PRACTICE EXERCISE

Complete a cost estimate based upon the following data. Overhead and profit will not be calculated in this example.

I. A landscape requires the following maintenance tasks and equipment:

a. 12,000 square feet of lawn cut 25 times each year with a power riding mower
b. 600 square feet of lawn cut 25 times each year with a 25-inch power hand mower
c. all lawn areas fertilized twice each year
d. all lawn areas raked once each spring with a 24-inch power rake
e. 700 square feet of shrub plantings cultivated with hoes twice each year
f. shrubs pruned once each year
g. shrubs fertilized once each year
h. 250 square feet of flower beds requiring soil conditioning once each spring
i. flowers in 250 square feet planted once each year
j. flowers in 250 square feet weeded by hand 10 times each year
k. flower beds in 250 square feet cleaned and prepared for winter once each autumn

II. Time requirements for the maintenance tasks include:

a. power riding mower cutting 1,000 square feet of lawn in 1 minute
b. the 25-inch power hand mower cutting 1,000 square feet of lawn in 3 minutes
c. lawn fertilization requiring 3 minutes per 1,000 square feet for spreading
d. lawn raking with a 24-inch power rake requiring 10 minutes per 1,000 square feet
e. hand hoe cultivation of the shrubs requiring 60 minutes per 1,000 square feet
f. pruning of shrubs averaging 60 minutes per 1,000 square feet
g. shrub fertilization requiring 5 minutes per 1,000 square feet
h. soil conditioning for flower beds requiring approximately 200 minutes per 1,000 square feet
i. flower planting requiring 600 minutes per 1,000 square feet
j. weeding of flowers requiring 60 minutes per 1,000 square feet
k. cleanup of flower beds in the fall requiring 400 minutes per 1,000 square feet

III. All laborers receive wages of $4.00 per hour.

IV. Material costs include:

a. lawn fertilizer at $4.00 per 1,000 square feet
b. shrub fertilizer at $2.00 per 1,000 square feet
c. conditioning materials for flower beds at $10.00 per 1,000 square feet
d. flowers for planting at $75.00 per 1,000 square feet

ACHIEVEMENT REVIEW

A. List three ways in which a cost estimate benefits a landscape maintenance firm.

B. What seven items of data are required before beginning a maintenance cost estimate?

C. Figure the total annual time in minutes for a task which is done 5 times a year and involves 6,000 square feet of area, and requires 7 minutes per 1,000 square feet to accomplish.

D. Figure the total material cost for mulch that is purchased to cover 1,500 square feet of area and costs $25.00 per 1,000 square feet. The mulch is applied once each year.

E. Calculate the total labor cost for a job that requires a total of 420 minutes. The worker assigned to the job is paid $4.00 per hour.

section 8

MAINTAINING THE LANDSCAPE

 # unit 25

PRUNING TREES AND SHRUBS

OBJECTIVES

After studying this unit, the student will be able to

- list the parts of a tree or shrub which are important to consider when pruning.
- explain how to determine when to prune, and why.
- describe which limbs are removed when pruning a tree and how it is done.
- describe which limbs are removed when pruning a shrub and how it is done.

Pruning is the removal of a portion of a plant to attain better shape or more fruitful growth. It is easily done, but not so easily done correctly. Each time that a bud or branch is removed from a plant, it creates both a short-term and long-term effect upon the plant. The *short-term effect* is how the plant looks immediately after pruning, and perhaps through the remainder of the current growing season. The *long-term effect* is how the plant appears after several seasons of growth without the part that has been pruned.

PARTS OF THE TREE

Before beginning to prune a plant, it is necessary to have a basic understanding of the anatomy of the plant. Figure 25-1 illustrates the parts of a tree.

The *lead branch* of a tree is the most important branch on the plant. It is dominant over the other branches, called the *scaffold branches.* The lead branch usually cannot be removed without losing the distinctive shape of the tree. This is especially true in young trees.

The scaffold branches create the *canopy* of the tree. The amount of shade cast by the canopy of a tree is directly related to the number of scaffold branches and the size of the leaves. When it becomes necessary to remove a branch from a tree, removal

Fig. 25-1 The parts of a tree

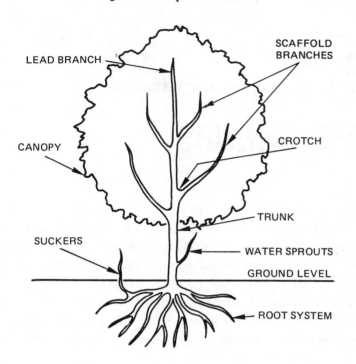

Fig. 25-3 The parts of a shrub

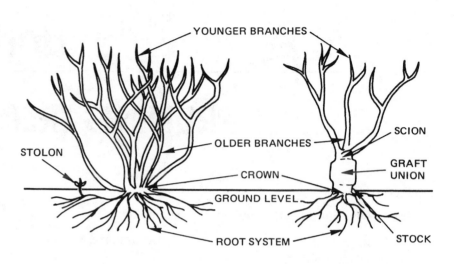

Fig. 25-2 Tree crotch structure

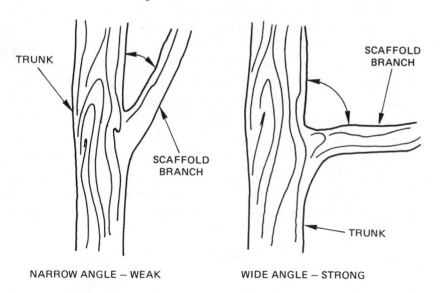

usually occurs at a *crotch,* that is, the point at which a branch meets the trunk of the tree or another, larger branch. It is always desirable to leave the strongest branches and remove the weakest. Where the crotch union is wide (approaching a right angle), the branch is strong. Where the crotch union is narrow, the branch is weak and may break in a heavy wind, figure 25-2.

Two other types of branches often found on trees are suckers and water sprouts. *Suckers* originate from the underground root system. *Water sprouts* develop along the trunk of the tree. Neither is desirable for a healthy tree and both should be removed.

PARTS OF THE SHRUB

A shrub is a multistemmed plant, figure 25-3. The stems of a shrub (called branches or twigs) differ in age within a single plant. The best flower and fruit production usually occurs on the younger branches. The younger branches are usually distinguished by a lighter color, less bark, and smaller diameter. The older branches have a darker color, are thicker in diameter, and possess a heavier bark.

The point at which the branches and the root system of a shrub meet is the *crown*. New branches originate at the crown, causing the shrub to grow wider. New shoots called *stolons* may spread underground from existing roots to create new shrubs from the parent plant. In cases where the shrub is the result of grafting

Fig. 25-4 The parts of a twig important to pruning

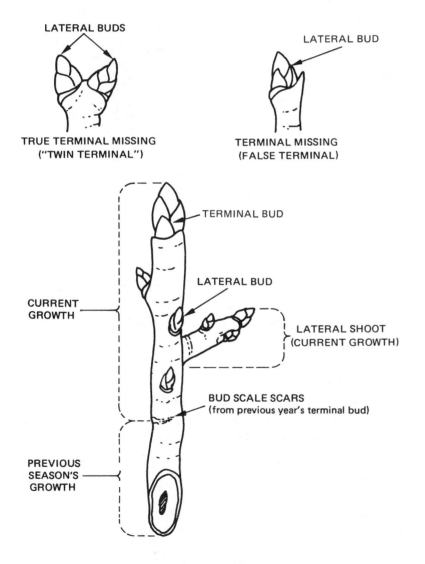

LATERAL BUDS

TRUE TERMINAL MISSING
("TWIN TERMINAL")

LATERAL BUD

TERMINAL MISSING
(FALSE TERMINAL)

TERMINAL BUD

LATERAL BUD

CURRENT
GROWTH

LATERAL SHOOT
(CURRENT GROWTH)

BUD SCALE SCARS
(from previous year's terminal bud)

PREVIOUS
SEASON'S
GROWTH

two plants together to make one, the *graft union* may also be seen at or near the crown. Graft unions are especially common in roses, but may exist on almost any ornamental shrub. Shoots originating below a graft union are from the *stock*, the root portion of the graft. They are cut away, since the quality of their flowers, fruit, and foliage is inferior. Only shoots originating from the *scion*, that portion of the graft occurring aboveground, are allowed to develop.

Even individual twigs have parts that are important to recognize when pruning. As illustrated in figure 25-4, each twig has a *terminal bud* and numerous *lateral buds*. The *bud* represents new growth for the plant. It may contain leaves, flowers, or both. The terminal leaf bud exerts dominance over the other buds. It gives the twig its added length each year. Should the terminal leaf bud be removed, the first lateral bud below it will eventually exert its dominance over the others and become the new terminal shoot.

Not all twigs produce their buds in the same arrangement. As figure 25-5 illustrates, lateral buds may be formed *opposite* each other, in an *alternate* fashion and, occasionally, in a *whorled* arrangement.

REASONS FOR PRUNING

Unfortunately, many people believe that trees and shrubs should be pruned simply because they have grown since the last pruning. This attitude tends to make the approach to pruning much like a haircut, resulting in a plant that is unnaturally shaped.

By understanding the reasons for pruning, the landscaper is better prepared to determine if a tree or shrub requires pruning. Pruning is done to

- control the size of the plant.
- improve the appearance of the plant by the removal of dead limbs or old wood.
- improve the health of the plant by the removal of diseased, weakened, or injured parts.

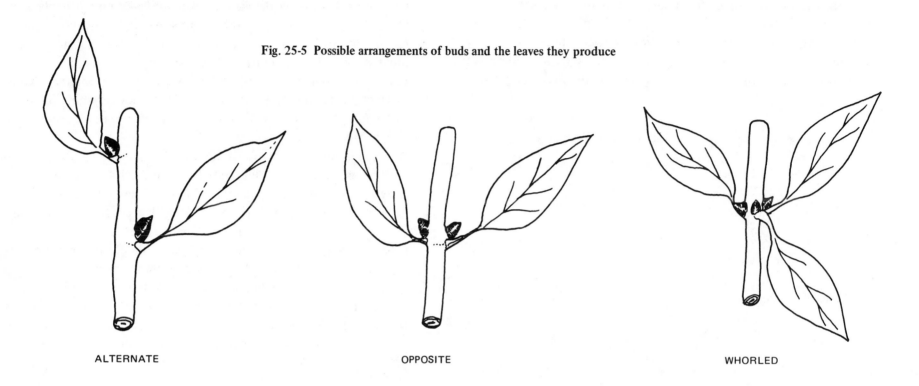

Fig. 25-5 Possible arrangements of buds and the leaves they produce

ALTERNATE   OPPOSITE   WHORLED

- train the plant to grow into a desired shape, such as with topiary pruning (geometric shaping) or espalier pruning (training plants to grow in a vinelike manner).

If a client requests maintenance pruning of plants and none of the above four reasons seem to apply, the landscaper should advise the client that pruning may be unnecessary.

THE PROPER TIME TO PRUNE

Landscapers who perform design and installation as well as maintenance work usually prefer to prune at times of the year when they have little other work. By doing this, their work and income are more evenly distributed throughout the year. Some plants can accept this off-season attention and remain unaffected by it. Other species accept pruning only during certain periods of the year.

There are advantages and disadvantages to pruning in every season. Since seasons vary greatly from region to region, the following information can be used only as a guide to pruning. Landscapers must determine the influence of local temperatures upon plant growth in their areas.

Winter Pruning

Winter pruning gives the landscaper off-season work. It also allows a view of the plant unblocked by foliage. Broken branches are easily seen, as are older and crossed branches. The major disadvantage of winter pruning is that without foliage, it is difficult to detect dead branches. Because of this, plants can become seriously misshapen if the wrong branches are removed. An additional disadvantage is the damage which can be done to frozen plant parts by cracking. Also, pruning scars have no opportunity to heal during winter, so the plant must carry the open wound throughout the season.

Summer Pruning

Summer pruning also provides off-season work for the landscaper. It allows time for all but very large wounds to heal before the arrival of winter. The major limitation of summer pruning is that problems of plants in full foliage may be concealed. Branches which should be removed are often difficult to see. Especially with trees, it is difficult to shape the branching pattern unless all of the limbs are visible.

Autumn Pruning

Autumn pruning can interfere with the landscaper's second busiest planting season. The income from planting is usually greater and faster than that from maintenance, so pruning in the fall is not welcomed by many landscapers. In terms of the health of the plant, autumn pruning is acceptable as long as it is done early enough to allow healing of the cuts prior to winter.

Autumn pruning should not be attempted on plants which bloom very early in the spring. These early bloomers produce their flower buds the preceding fall. Thus, fall pruning cuts away the flower buds and destroys the spring color show. Autumn pruning should be reserved for those plants which bloom in late spring or summer. They produce their flower buds in the spring of the year in which they bloom. There is no danger of cutting away buds in the fall since there are none present.

Spring Pruning

Spring pruning is usually not welcomed by the landscaper, since spring is the major planting season. However, most plants can be successfully pruned early in the spring as buds begin to swell. This permits a clear view of the live and dead branches. There is not yet any foliage to block the view of the complete plant. If the plant is an early spring bloomer, it is best to prune it immediately after flowering. Spring pruning also provides the plant with time to heal any wounds. Likewise, the unfolding leaves conceal the fresh cuts from the viewer's eye.

There are some exceptions to the guidelines for spring pruning; these are the needled evergreens and any plants which bleed severely when cut in the spring. While needled evergreens or holly can be pruned at any season, people often want to use the cut greens as decoration during the winter holiday season; thus, it is best to prune at this time. When removing branches, care should be taken not to break frozen limbs and twigs.

Also requiring some special attention are plants which have high sap pressure in the early spring. These varieties should not be pruned until the pressure has been reduced naturally by the onset of summer or fall. These are better times to prune such plants as maples, birches, walnuts, or poinsettias since they will bleed rather than heal quickly after cutting in the spring.

The Parts of the Plant to Prune

The actual limbs and branches that are removed from a tree or shrub are determined by the reason for the pruning. If the objective is to remove diseased portions, the cut should be made through healthy wood between the trunk or crown and the infected part. The cut should never be made through the diseased wood or just behind it. If this is done, the pruning tool becomes contaminated and may transmit the disease-causing organism to healthy parts that may be pruned later, figure 25-6.

If the pruning is being done for the overall health and appearance of the plant, those branches which are growing into the center of the plant are removed. Limbs and twigs which are growing across other branches can crowd and harm the plant. Such branches should be selected for removal. If more than one limb originates at a tree crotch, the strongest should be left and the others removed. Figure 25-7 illustrates these pruning needs.

If the pruning is being done to reduce the size of the plant without altering its natural shape, careful selection of both the pruned branches and those allowed to remain must be made.

Fig. 25-6 The correct way to remove diseased limbs or twigs

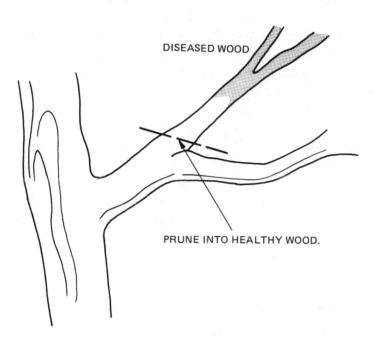

DISEASED WOOD

PRUNE INTO HEALTHY WOOD.

Fig. 25-7 Which branches to prune for improved plant appearance

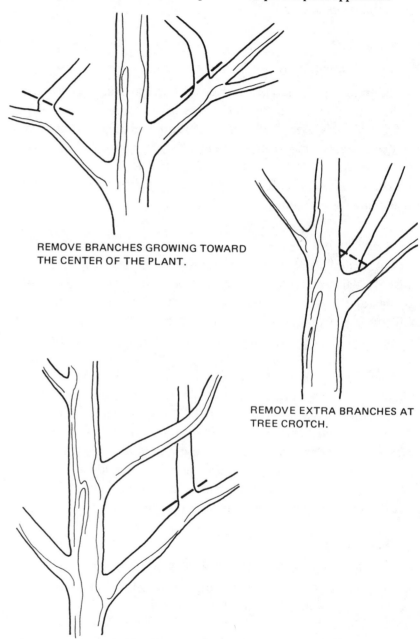

REMOVE BRANCHES GROWING TOWARD
THE CENTER OF THE PLANT.

REMOVE EXTRA BRANCHES AT
TREE CROTCH.

REMOVE BRANCHES GROWING ACROSS
OTHER BRANCHES.

Major structural limbs and twigs must be left so that no holes appear in the plant. The fact that many secondary branches stem from one older branch is often overlooked. The result is that removal of the older branch creates a tree or shrub with an entire side missing.

If, as with evergreens, the purpose of the pruning is to create a denser foliage, it is the center shoot which is shortened or removed, figure 25-8. Doing so encourages the lateral buds to grow and create several shoots, where formerly there had been only one. The result is added plant fullness.

HOW TO PRUNE

Pruning Tools

Unit 3 introduced some of the tools used in pruning. A brief review is provided here to aid the student in proper tool selection — an essential step in correct pruning methods. Using the proper

Fig. 25-8 Evergreens are pruned in the spring if a denser foliage is desired.

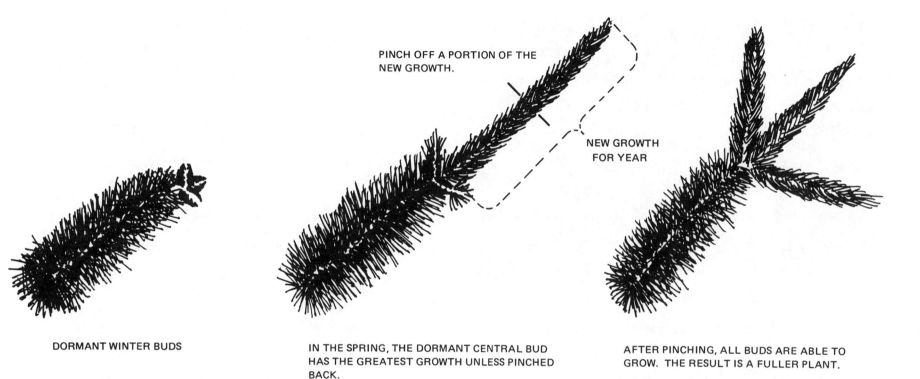

PINCH OFF A PORTION OF THE
NEW GROWTH.

NEW GROWTH
FOR YEAR

DORMANT WINTER BUDS

IN THE SPRING, THE DORMANT CENTRAL BUD
HAS THE GREATEST GROWTH UNLESS PINCHED
BACK.

AFTER PINCHING, ALL BUDS ARE ABLE TO
GROW. THE RESULT IS A FULLER PLANT.

tool protects the plant against damage which can result from incorrect tool usage.

The *hand pruner,* figure 25-9, is used to cut branches of up to about 1/2 inch in diameter. It is available in a wide range of prices, with the higher priced tools made of the best steel and having the most durable parts.

Lopping shears, figure 25-10, are used for the removal of branches that would cause a strain on hand pruners. They are effective on wood 1 to 1 1/2 inches in diameter.

The *pruning saw* is needed for tree limbs and shrub wood which exceed 1 1/2 inches in diameter, figure 25-11. It is available with a single blade or a double blade.

Fig. 25-9 Hand pruners are used to cut woody stems of up to 1/2 inch in diameter.

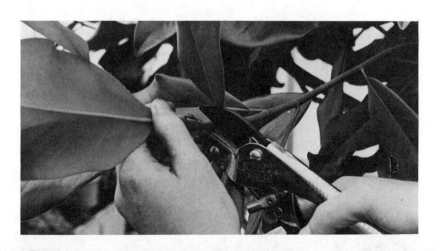

Fig. 25-10 Lopping shears are used for removing branches measuring up to 1 1/2 inches in diameter.

Pruning Methods

The method of pruning a tree or shrub depends upon the size and amount of branches being removed.

If a limb is being removed from a tree with a pruning saw, the technique is called *jump-cutting*. This method allows the scaffold limb to be removed without stripping off a long slice of bark with it as it falls. The jump-cut requires three cuts for safe removal of a limb, figure 25-12. The final cut should remove the stub of the limb as close to the trunk as possible. The wound is then covered with a wound paint to seal it from insects and disease until the plant has time to heal. Large wounds may require several applications of wound paint until healed.

Fig. 25-12 The removal of large limbs using the technique of jump-cutting. The cut at (A) allows the limb to snap off after a cut at (B) without stripping bark from the trunk as it falls. The final cut at (C) removes the stub.

Fig. 25-11 Larger limbs are removed with a pruning saw.

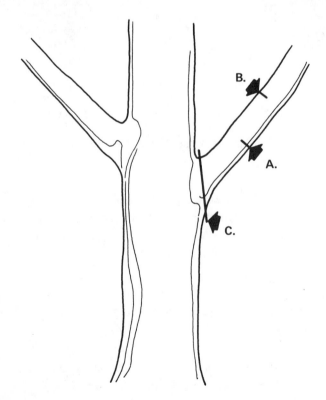

When shrubs are pruned, one of two techniques is used. *Thinning out* is the removal of a shrub branch at or near the crown. It is the major means of removing old wood from a shrub while retaining the desired shape and size. *Heading back* is the shortening, rather than total removal, of a twig. It is a means of reducing the size of the shrub, figure 25-13. In cases where shrubs have become tall and sparse, a combination of thinning out and heading back can rejuvenate an old planting, figure 25-14.

In heading back, the twig is shortened, but not completely removed. The location of the cut is not simply left to chance. Figure 25-15 illustrates three possible placements for twig removal. Twig A has too much wood remaining above the bud. It will die from the point of the cut back to the bud, but may not heal over quickly enough to prevent insect and disease entry. The woody stub itself may also decay later. Twig B is cut below the bud, causing the bud to dry out and possibly die. Twig C is pruned

Fig. 25-13 The techniques of thinning out and heading back

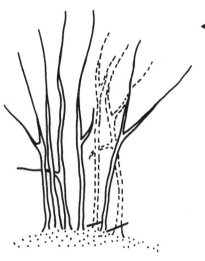

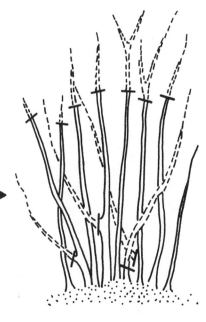

THINNING OUT. As its name implies, this method involves selection of an appropriate number of strong, well-located stems and removal at the ground level of all others. This is the preferred method for keeping shrubs open and in their desired shrub size and form. With most shrubs, it is an annual task; with others, it is required twice a year.

HEADING BACK. This method involves trimming back terminal growth to maintain desired shrub size and form. It encourages more compact foliage development by allowing development of lateral growth. This is the preferred method for controlling the size and shape of shrubs and for maintaining hedges.

Fig. 25-14 Two techniques used to rejuvenate old shrubs

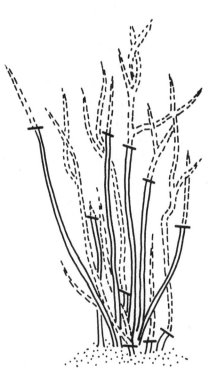

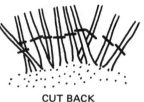

CUT BACK

SELECT SIX OR MORE WELL-PLACED VIGOROUS SHOOTS

HEAD BACK

GRADUAL RENEWAL. This pruning method involves removal of all mature wood over a 3-to 5-year period. Approximately one-third of the mature wood is removed each season. This is the preferred method for shrubs that have not been recently pruned and are somewhat overgrown.

COMPLETE RENEWAL. This method involves complete removal of all stems at the crown or ground level. Two to three months later the suckers or new growth that emerges is thinned to the desired number of stems. These, in turn, are headed back to encourage lateral branching. Unpruned, seriously overgrown, or severely damaged shrubs are prime prospects for this treatment.

correctly. It is cut just above the bud and parallel to the direction in which the bud is pointing. The cut is close enough to the living tissue to heal over quickly. However, it is not so close to the bud that it promotes drying.

The direction in which the bud grows can be guided by good pruning techniques. Since branches growing into the plant can create congestion, they may be discouraged by the selection of an outward-pointing bud when heading back, figure 25-16. If the twig has buds pointing in opposite directions, the unnecessary one is removed.

Whenever the pruning cut exceeds 1 inch in diameter, wound paint may be applied to protect the plant from infection during the healing process. Wound paints are available in aerosol cans for small-scale use and in large sizes for professional use. Despite the claims of manufacturers that some offer greater protection than others, there is little evidence to suggest much difference in quality among brands. Some recent research has suggested that wound paints may delay the healing of the plant tissue. Where this is suspected, the landscaper may choose not to use wound paint.

Fig. 25-15 Where to prune the twig

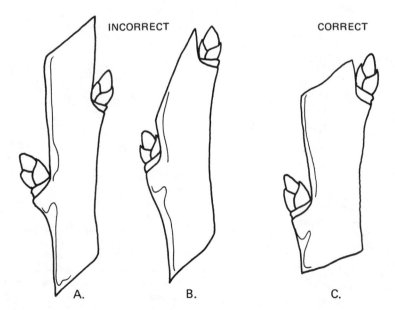

Fig. 25-16 Twigs should be pruned to leave an outward pointing bud

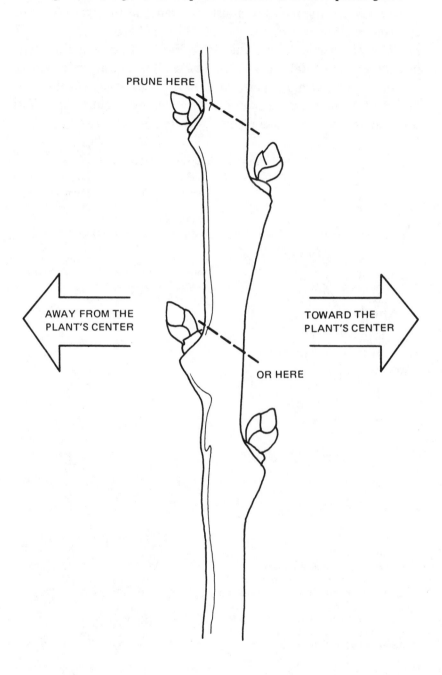

HOW TO PRUNE HEDGES

Creation of a hedge requires close spacing of the shrubs at the time of planting and a special type of pruning. The pruner must shear the plant so that it is as dense as possible. This is usually done with hedge shears, figure 25-17. Hedge shears easily cut through the soft new growth of spring, the time at which most hedges are pruned. For especially large hedges, electric shears are available. However, practice and skill are required for the use of electric shears. Damage occurs quickly if the landscaper does not keep the shears under control.

Proper pruning of a hedge requires that it not only be level on top, but tapered on the sides. It is important that sunlight be able to reach the lower portion of the hedge if it is to stay full. When sunlight cannot reach the lower parts of the hedge, it becomes leggy and top-heavy in appearance. Figure 25-18 illustrates correct and incorrect forms of hedge pruning.

Fig. 25-17 Hedge shears

Fig. 25-18 Correct and incorrect hedge forms

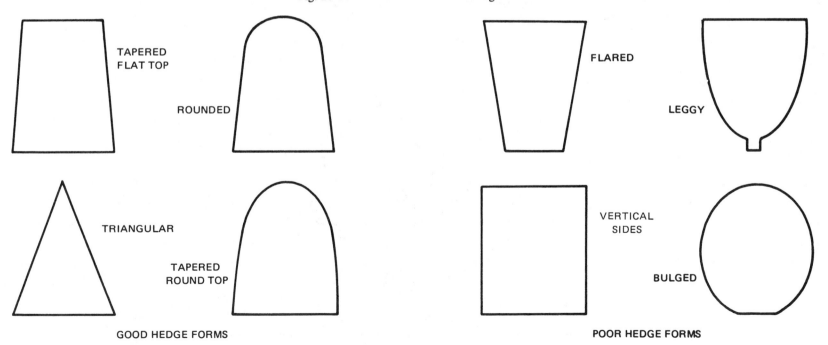

TAPERED FLAT TOP

ROUNDED

TRIANGULAR

TAPERED ROUND TOP

GOOD HEDGE FORMS

FLARED

LEGGY

VERTICAL SIDES

BULGED

POOR HEDGE FORMS

ACHIEVEMENT REVIEW

A. Match each term in column 1 with the correct definition from column 2.

Column 1

a. thinning out
b. heading back
c. lateral bud
d. terminal bud
e. graft union
f. tree
g. shrub
h. jump-cut
i. crown

Column 2

1. the end bud on a branch
2. a plant having a single, dominant central trunk
3. the complete removal of a shrub branch at the base of the plant
4. the point of a shrub at which branches and roots meet
5. a multistemmed plant with no central trunk
6. a technique for removal of large tree limbs
7. the side bud on a branch
8. the shortening of a shrub branch
9. the junction between a stock and a scion

B. 1. Label the parts of the tree indicated in the following drawing.

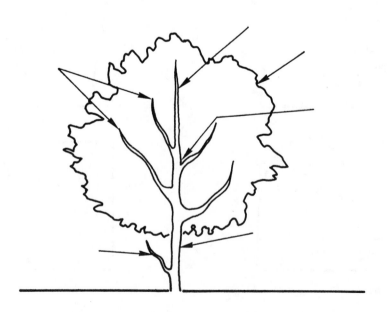

2. The following drawing shows seven numbered branches in the tree. Which three branches should be removed, and why?

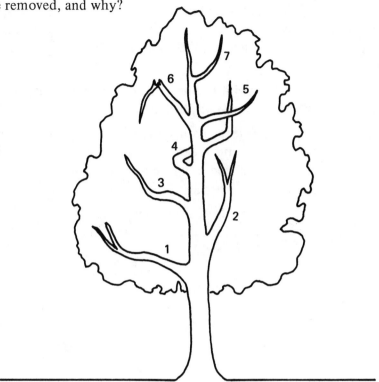

3. Indicate the three cuts needed for a correct jump-cut removal of the limb shown in the following drawing. Number the cuts in the order in which they would be made.

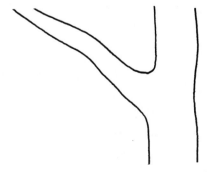

SUGGESTED ACTIVITIES

1. Plan a pruning day. Each student should borrow a pair of hand pruners, lopping shears, or a pruning saw. Try for an assortment of plants needing pruning, such as overgrown, leggy, and sparse. Work in pairs and proceed slowly, discussing major decisions about where to prune.

2. Collect a variety of budded twigs having both opposite and alternate bud arrangements. Practice the correct heading back technique.

3. For a winter activity, select several mature deciduous trees and shrubs for analysis. Study the overall branching patterns. Then discuss which limbs of the trees should be removed; which shrubs need old wood removed; which branches are important to the shape of the plant; and which branches are dispensable.

4. Invite a guest speaker to the class to explain topiary and espalier pruning techniques. Local college instructors, Cooperative Extension specialists, or area garden clubs may be able to furnish names of nearby experts.

unit 26

CARE OF THE LAWN

OBJECTIVES

After studying this unit, the student will be able to

- outline a maintenance program for professional lawn care.
- explain the meaning of fertilizer analysis statements.
- determine the amount of fertilizer needed to cover a specific lawn.
- compare the maintenance requirements of the various turf grasses.

Of all aspects of the landscape which require maintenance during the year, lawns consume the most time. A lawn necessitates both seasonal care and weekly care. Like any other feature of the landscape, it is easier to maintain if it is installed properly. Thus, the landscaper who is hired to install and maintain a lawn may have an easier job than the landscaper hired to maintain a lawn which was poorly installed by someone else.

SPRING LAWN CARE

Cleanup

Spring operations begin the season of maintenance. In areas of the country where winters are long and hard, lawns may be covered with compacted leaves, litter, or semidecomposed thatch. The receding winter may leave behind grass damaged by salt injury, disease, or freezing and thawing.

Small areas can be cleaned of debris with a strong rake. Larger areas of several acres or more require the use of such equipment as power sweepers and thatch removers to accomplish the same type of cleanup.

Rolling of the Lawn

In many central and northern states, the ground freezes and thaws many times during the winter. Such action can cause *heaving* of the turf grass. Heaving pulls the grass roots away from the soil, leaving them exposed to the drying wind. Heaving also creates a lumpy lawn.

Where heaving occurs, it is advisable to give the lawn a light rolling with a lawn roller in the spring. Rolling presses the heaved turf back in contact with the soil. The roller is applied in a single direction across the lawn, followed by a second rolling at an angle perpendicular to the first.

Two precautions should be observed before rolling a lawn. One is that clay soil should never be rolled, since air can be easily driven from a clay soil and the surface quickly compacted. The other precaution is that no soil should be rolled while wet. The roller can be safely used only after the soil has dried and regained its firmness.

The First Cutting

The first cutting of the lawn each spring removes more grass than the cuttings which follow later in the summer. The initial cutting at 1/2 to 3/4 inch is done to promote horizontal spreading of the grass. This, in turn, hastens the thickening of the lawn. An additional benefit of the short first cutting is that fertilizer, grass seed, and weed killer which are applied to the lawn reach the soil's surface more easily. Cuttings done later in the year are usually not as short.

Reseeding

Reseeding may be needed if patches of turf have been killed by disease, insects, or other causes. Widespread thinness of the grass does not indicate a need for reseeding. It indicates a lack of fertilization or improper mowing.

Reseeding is warranted where bare areas are at least 1 foot in diameter. Seed should be selected to match the grasses of the established lawn. The surface of the soil can be broken with a rake and a mixture of seed and topsoil applied over it, figure 26-1. (A good mixture for reseeding is 1 pound of grass seed to every bushel of topsoil.) The soil should be kept moist for about three weeks to assure successful germination of the seed.

The timetable for reseeding is the same as that for seeding, as described in Unit 22. It is specifically related to the type of grass involved (warm-season or cool-season).

Aeration

Aeration of a lawn is the addition of air to the soil. The presence of air in the soil is essential to good plant growth. If the lawn is installed properly, the incorporation of sand and organic material into the soil promotes proper aeration. However, where traffic is heavy or the clay content is high, the soil may become compacted. The grounds keeper can relieve the compaction by use of a power aerator, figure 26-2. There are several types of aerators. All cut into the soil to a depth of about 3 inches and remove plugs of soil or slice it into thin strips.

Following the use of the aerator, a covering of organic material is applied to the lawn. Running a rotary power mower

Fig. 26-1 Patch seeding of thin areas in established lawns is done by breaking the soil surface and applying a small handful of seed.

Fig. 26-2 The aerator is used to remove plugs of soil from compacted lawns, allowing air to enter the soil.

over the organic material blows it into the slits or holes left by the aerator.

LAWN FERTILIZATION

Much like grass seed, lawn fertilizer is sold in an assortment of sizes and formulations and priced accordingly. Stores selling fertilizers range from garden centers to supermarkets. The professional grounds keeper needs to have a basic knowledge of fertilizer products prior to their purchase. Otherwise, it is difficult to choose among the many brands available.

Nutrient Analysis and Ratio

The fertilizer bag identifies its contents. It displays three numbers which indicate its *analysis,* that is, the proportion in which each of three standard ingredients is present. These numbers, such as 10-6-4 or 5-10-10, indicate the percentage of total nitrogen, available phosphoric acid, and water-soluble potash present in the fertilizer, figure 26-3. The numbers are always given in the same order and always represent the same nutrients.

With simple arithmetic, fertilizers can be compared on the basis of their *nutrient ratio*. For example, a 5-10-10 analysis has a ratio of 1-2-2. (Each of the numbers has been reduced by dividing by a common factor, in this case, 5.) A fertilizer analysis of 10-20-20 also has a ratio of 1-2-2. Therefore, a 5-10-10 fertilizer supplies the three major nutrients in the same proportion as a 10-20-20 fertilizer, but twice as much of the actual product must be applied to obtain the same amount of nutrients as is contained in the 10-20-20 fertilizer.

Fig. 26-3 How to interpret fertilizer analysis figures. The nutrients are always shown in the same order.

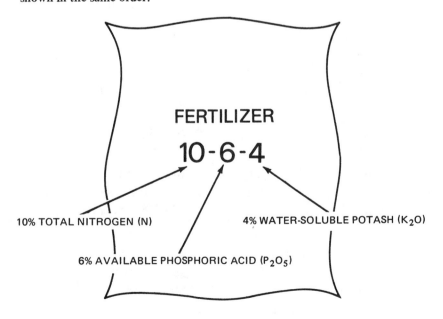

FERTILIZER

10-6-4

10% TOTAL NITROGEN (N)   4% WATER-SOLUBLE POTASH (K_2O)

6% AVAILABLE PHOSPHORIC ACID (P_2O_5)

Example:

50 pounds of 5-10-10 fertilizer contain:	50 pounds of 10-20-20 fertilizer contain:
2 ½ pounds of N (nitrogen)	5 pounds of N
5 pounds of P_2O_5 (phosphoric acid)	10 pounds of P_2O_5
5 pounds of K_2O (potash)	10 pounds of K_2O

The ratio of the fertilizers is the same, but the amount of nutrients available in a bag of each differs. The 5-10-10 mixture should be less expensive than the higher analysis material.

Thus, one measure of the quality of a fertilizer is its analysis. The higher the analysis, the greater is the cost. Whether or not a high analysis fertilizer is needed must be determined by the individual plant. Generally, residential lawns do not need a high analysis fertilizer.

Forms of Nitrogen Content

Another factor influencing the quality and cost of fertilizers is the form of nitrogen they contain. Some fertilizers contain nitrogen in an *organic form*. Examples include peat moss, peanut hulls, dried blood, tobacco stems, sewage sludge, and cottonseed meal. The nitrogen content of these materials ranges from 1 1/2 to 12 percent, depending upon the particular material. While sewage sludge is used to some extent on golf course turf, organic fertilizers are not widely used for fertilization of grasses because they are too low in nitrogen. Often, the nitrogen which is present is not in a form which can be used by plants. The best use of organic fertilizers is as soil conditioners which greatly improve the water retention and aeration of the soil.

Chemical forms are the most commonly used fertilizers. They contain a higher percentage of nitrogen. The nitrogen may be quickly available or slowly available; this determines the timing of the nitrogen's release into the soil and uptake by the grass or other plants. It also influences the cost of the fertilizer.

Quickly available fertilizers usually contain water-soluble forms of nitrogen. This means that the nitrogen can be *leached* (washed) through the soil before the plants take it in through their root systems. *Slowly available* fertilizers (also called *slow-release*) make their nitrogen available to the plant more gradually and over a longer period of time. The slow-release effect is possible because the nitrogen used is in a form that is insoluble in water. This gives the plants more time to absorb the nitrogen and prevents fertilizer burn of the plant. Slow-release fertilizers are therefore more expensive than the quickly available forms. Slow-release fertilizers

are usually labeled as such. This helps the landscaper to know what is being purchased and what to expect as a response from the plants.

Fillers

A final factor affecting the price and quality of a fertilizer is the amount of filler material it contains. This is directly related to the analysis of the product. In addition to the three major nutrients, fertilizers may contain additional *trace elements* (nutrients which are essential, but needed in smaller amounts) and filler material. *Filler material* is used to dilute and mix the fertilizer. Certain fillers also improve the physical condition of mixtures. However, filler material adds weight and bulk to the fertilizer, thereby requiring more storage space.

The following listing compares high analysis fertilizers (those with a high percentage of major nutrients) and low analysis fertilizers (those with a low percentage of major nutrients) on various points.

High Analysis Fertilizer	Low Analysis Fertilizer
• Contains more nutrients and less filler	• Contains fewer nutrients and more filler
• Cost per pound of actual nutrients is less	• Cost per pound of actual nutrients is greater
• Weighs less; less labor is required in handling	• Is bulky and heavy; more labor is required in handling
• Requires less storage space	• Requires more storage space
• Requires less material to provide a given amount of nutrients per square foot	• Requires more material to provide a given amount of nutrients per square foot
• Requires less time to apply a given amount of nutrients	• Requires more time to apply a given amount of nutrients

In summary, fertilizer cost is determined by three major factors: analysis, form of nitrogen, and amount of bulk filler material. The higher the analysis and the greater the percentage of slow-release nitrogen, the more expensive is the fertilizer.

When to Fertilize Lawns

Lawns should be fertilized before they need the nutrients for their best growth. Cool-season grasses derive little benefit from fertilizer applied at the beginning of the hot summer months; only the weeds benefit from nutrients applied during the late spring. Cool-season grasses should be fertilized in the early spring and early fall. This supplies proper nutrition prior to the seasons of greatest growth. Landscapers should never practice late fall fertilization; it encourages soft, lush growth which is damaged severely during the winter.

Warm-season grasses should receive their heaviest fertilization in late spring. Their season of greatest growth is the summer.

Amount of Fertilizer

The amount of fertilizer to use is usually stated in terms of the number of pounds of nitrogen to apply per 1,000 square feet. The number of pounds of nitrogen in a fertilizer is determined by multiplying the weight of the fertilizer by the percentage of nitrogen it contains.

Examples

Problem: How many pounds of actual nitrogen are contained in a 100-pound bag of 20-10-5 fertilizer?

Solution: 100 pounds x 20% N = pounds of N

100 x 0.20 = 20 pounds of N

Problem: How many pounds of 20-10-5 fertilizer should be purchased to apply 4 pounds of actual nitrogen to 1,000 square feet of lawn?

Solution: Divide the percentage of N into the pounds of N desired. The result is the number of pounds of fertilizer required.

4 pounds of N desired ÷ 20% = pounds of fertilizer required

4 ÷ 0.20 = 20 pounds of 20-10-5 fertilizer required

The following table lists general fertilizer recommendations for various grasses. It also states the pounds of fertilizer needed to provide essential nutrients to 1,000 square feet of lawn.

Grass	Pounds of Nitrogen Needed Per 1,000 Sq. Ft.	Pounds of Common Fertilizers Needed Per 1,000 Sq. Ft.			
		20-10-5	10-10-10	12-4-8	5-10-5
Kentucky bluegrass	4	20	40	33	80
Merion bluegrass	5 to 7	25 to 35	50 to 70	42 to 58	100 to 140
Bentgrass	5 to 7	25 to 35	50 to 70	42 to 58	100 to 140
Fescues	4	20	40	33	80
Bermuda grass	4 to 9	20 to 45	40 to 90	33 to 75	80 to 180
Bahia	1 to 3	5 to 15	10 to 30	8 to 24	20 to 60
Carpet grass	1 to 3	5 to 15	10 to 30	8 to 24	20 to 60
Centipede grass	1 to 3	5 to 15	10 to 30	8 to 24	20 to 60
St. Augustine grass	2 to 5	10 to 25	20 to 50	16 to 42	40 to 100
Zoysia grass	2 to 5	10 to 25	20 to 50	16 to 42	40 to 100

When applying fertilizer to lawns, the recommended poundage should be divided into two or three applications. For example, the 4 pounds of nitrogen per 1,000 square feet for bluegrasses and fescues might be applied at the rate of 2 pounds in the early spring and 2 pounds in the early fall. Another possibility is to apply 1 pound in early spring, 1 pound in midsummer, and 2 pounds in early fall. A spreader must be used to assure even distribution of the fertilizer. It is applied in two directions with the rows slightly overlapped, figure 26-4.

WATERING THE LAWN

Turf grasses are among the first plants to show the effects of lack of water, since they are naturally shallow rooted as compared to trees or shrubs. The grounds keeper should encourage deep root growth by watering so that moisture penetrates to a depth of 8 to 12 inches into the soil. Failure to apply enough water so that it filters deeply into the soil promotes shallow root growth, figure 26-5. Such shallow root systems can be severely injured during hot, dry summer weather.

For water to penetrate sufficiently, 1 to 1 1/2 inches of water must be applied at each watering. The amount of water given off by a sprinkler can be calibrated once, with the setting noted for later reference.

To calibrate a sprinkler, set several wide-topped, flat-bottomed cans with straight sides (such as coffee cans) in a straight line out from the sprinkler. When most of them contain 1 to 1 1/2 inches of water, shut off the sprinkler. The amount of time required should be noted for future use. Figures 26-6 and 26-7 illustrate two types of portable sprinklers. In addition, permanently installed irrigation systems are available (at considerable cost) for large turf plantings.

The best time of day to water lawns is between early morning and late afternoon. Watering in the early evening or later should be avoided because of the danger of disease; turf diseases thrive in lawns that remain wet into the evening. Watering prior to evening gives the lawn time to dry before the sun sets.

Fig. 26-4 Distributing fertilizer by use of a spreader

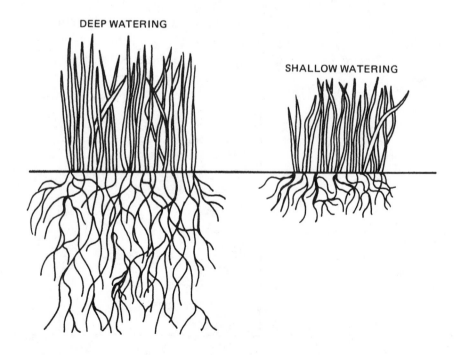

Fig. 26-5 Deep watering promotes deep, healthy root growth. Shallow watering promotes shallow rooting and leaves the grass susceptible to injury by drought.

DEEP WATERING

SHALLOW WATERING

Fig. 26-6 One type of lawn sprinkler. An arch of water is cast from side to side. This sprinkler requires periodic relocation.

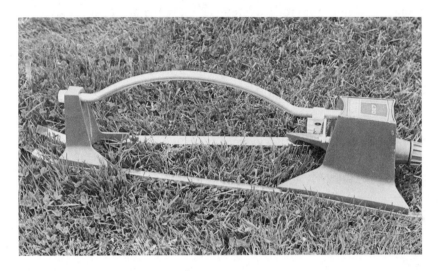

If watering is done at the proper time and to the proper depth, it is necessary only once or twice each week. Infrequent, deep watering is much better than daily, shallow watering.

MOWING THE LAWN

There are two types of mowers available for the maintenance of lawns: the reel mower, figure 26-8, and the rotary mower, figure 26-9. While reel mowers give a finer cut, the rotary mower is more popular, probably because it lends itself more easily to powered drive.

Not all species of grass are mowed to the same height. Therefore, when mixing species (as in a seed blend), it is important that all species in the combination have the same recommended cutting

Fig. 26-7 Another type of lawn sprinkler. The water is thrown in a circular pattern. The mechanism slowly pulls itself across the lawn, requiring no relocation by the grounds keeper.

Fig. 26-8 Reel mower. The blades move parallel with the wheels and drop down to cut off the grass.

Fig. 26-9 Rotary mower. The blades move perpendicular to the wheels. They cut across the lawn's surface to remove the grass.

height. The following is a list of the recommended heights for some of the more common lawn grasses.

Grass	Mowing Height
Bahia grass	2 ½ to 3 inches
Bentgrass	½ to ¾ inch
Bermuda grass	½ to 1 inch
Bluegrasses	1 ½ to 2 inches
Carpet grass	1 ¼ to 2 inches
Centipede grass	1 ½ to 2 inches
Fescues	1 ½ to 2 inches
Ryegrasses	1 ½ to 2 inches
St. Augustine grass	1 ½ to 2 ½ inches
Zoysia grass	½ to 1 ¼ inches

During very hot summer weather, cool-season grasses benefit if left an additional half inch of height. The added height helps to shade the root system and prevents the onset of dormancy and browning.

CONTROLLING WEEDS IN LAWNS

There are many definitions of the word *weed*; student landscapers undoubtedly have their own definitions. For discussion purposes, a *weed* may be defined as a plant which (1) is growing where it is not wanted, and (2) has no apparent economic value. Beyond these two qualities, weeds have little else in common. Some weeds have broad leaves (such as dandelion, plantain, and thistle); others are grasses. Some are annuals; others are perennials.

In lawns, weeds are professionally controlled almost entirely by chemical means. A chemical which kills weeds is called a *herbicide.* Some herbicides are *selective,* meaning that they kill only certain weeds and do not harm other plants. Other herbicides are *nonselective;* these kill any plant with which they come into contact. Certain herbicides are classified as *preemergence* types. They are applied before the weed seeds *germinate* (sprout) and kill them as they sprout. *Postemergence* herbicides are applied to weeds after they have germinated.

Because the regulations governing the purchase and use of herbicides and other pesticides vary among states, this text contains no recommendations of specific weed controls. Professional grounds keepers in all states can obtain current recommendations for pesticides approved for use in their state by consulting a college of agriculture or the Cooperative Extension Service. The current trend is toward registration of pesticide applicators and restriction of the more poisonous pesticides to use by professionals only. Herbicides are deadly poisons, and the landscaper should handle them cautiously and with a serious attitude.

Good landscaping practices are the best defense against weeds. However, the need for some herbicides is almost certain to exist in most lawns. They are available in liquid or powdered form. Sometimes herbicides are mixed with fertilizer and sold as combination products which save time and labor.

If an herbicide is applied in liquid form, the sprayer should be cleaned afterward and set aside to be used only for that purpose. Small amounts of herbicide which may remain in the sprayer can kill valuable ornamental plants if the landscaper uses the same sprayer to apply another liquid material directly to plants.

The following chart summarizes the characteristics of the most common grasses and indicates their basic maintenance requirements.

SUMMARY OF TURF GRASS CHARACTERISTICS AND MAINTENANCE

Grass Type	Texture	Mowing Frequency	Number of Fertilizer Applications Per Yr.	Preferred Soil Type	Light Preference	Wear Resistance	Method of Planting	Mowing Height, in Inches
Bahia grass	medium to coarse	weekly	1 or 2	acidic	semishade	good	seed or plugging	2 ½ to 3
Bentgrass	fine	once or twice each week	3	neutral to slightly acidic	sun or semishade	good	seed	½ to ¾
Bermuda grass	fine	1 to 3 times weekly	4 to 12	wide range	sun	good	seed or plugging	½ to 1
Bluegrass	medium	weekly	2 to 3	neutral to slightly acidic	sun	good	seed	1 ½ to 2
Carpet grass	medium	weekly	1	wet, poorly drained, and acidic	sun or semishade	poor	seed or plugging	1 ¼ to 2
Centipede grass	medium	every two weeks	1	acidic	sun or semishade	poor	seed or plugging	1 ½ to 2
Fescues	fine	weekly	3	neutral to slightly acidic	semishade	good	seed	1 ½ to 2
Ryegrasses	medium to coarse	weekly	3	wide range	sun or semishade	good	seed	1 ½ to 2
St. Augustine grass	medium to coarse	weekly	3 to 4	wide range	sun or shade	good	plugging	½ to 2 ½
Zoysia grass	fine to medium	2 to 4 times monthly	3 to 4	wide range	sun or semishade	good	plugging	½ to 1 ¼

ACHIEVEMENT REVIEW

A.   Define the following terms.

1. heaving   3. fertilizer analysis   5. slow-release fertilizer
2. aeration   4. low analysis fertilizer   6. herbicide

B.   What does *10-6-4* on a bag of fertilizer mean?

C.   Would 10-6-4 fertilizer be considered a high analysis or low analysis product? Why?

D.   Of the three fertilizers listed below, which two have the same ratio of nutrients?

    10-20-10   5-10-5   5-10-15

E.   Why might the prices of two 50-pound bags of fertilizer differ greatly?

F.   At what time of the year are warm-season grasses fertilized? Cool-season grasses?

G.   How many pounds of actual nitrogen are contained in a 50-pound bag of 12-4-8 fertilizer?

H.   How many pounds of 10-10-10 fertilizer should be purchased to fertilize a 2,000 square-foot lawn of bluegrass and fescue if it is applied at the recommended rate?

I.   How much water is needed to deeply soak an average lawn?

J.   What is the best time of day to water lawns?

SUGGESTED ACTIVITIES

1.   Study the chart at the end of the unit and match grasses grown in the local area which could be successfully blended and grown together. Compare grasses on the basis of texture, frequency of fertilization, mowing height, and preferred soil type.

2.   Choose a partner and form a team of grounds keepers. As a team, outline a maintenance program for a lawn in the local area measuring 10,000 square feet, using grasses common to the area. Assume that the soil has a pH of 5.5 and is heavy in clay and low in nutrients, and that the grass is sparse. The lawn is partially shaded with trees. The program outline should include seasonal and weekly operations, an itemization of equipment needed, and specific amounts of materials required.

3.   Calibrate a sprinkler, following the directions given in this unit. Determine the length of time required to apply 1 inch of water.

4.   Measure the thatch layer of several lawns near the school. Select one lawn where clippings are collected when the lawn is mowed and another where they are not collected. The thatch layer should be thickest on the lawn where clippings are allowed to remain on the lawn.

5.   Visit a lawn equipment dealership. Ask the owner to show the various models of mowers, spreaders, sprayers, rollers, and rakes that are stocked for lawn maintenance.

WINTERIZATION OF THE LANDSCAPE

OBJECTIVES

After studying this unit, the student will be able to

- list those elements of the landscape which require winter protection.
- describe eight possible types of winter injury.
- explain nine ways to protect against winter injury.

Winter injury is any damage done to elements of the landscape during the cold weather season of the year. The injury may be due to natural causes or to human error. It may be predictable or totally unexpected. At times winter injury can be avoided, while at other times it can only be accepted and dealt with.

Winter injury attacks most elements of the outdoor room. Plants, paving, steps, furnishings, and plumbing are all susceptible to damage from one or more causes.

TYPES OF WINTER INJURY

While the types of winter damage are almost unlimited, there are several which commonly occur. The landscaper should be especially aware of these. Injuries are caused by one of two agents: nature or human beings. There are many different examples within these two general categories.

Natural Injuries

The severity of winter weather can cause extensive damage to plant materials in the landscape.

Windburn results when evergreens are exposed to strong prevailing winds throughout the winter months. The wind dries out the leaf tissue, and the dehydrated material dies. Windburn causes a brown to black discoloration of the leaves on the windward side of the plant. Very often, leaves further into the plant or on the side opposite the wind show no damage, figure 27-1. Broad-leaved evergreens are the most susceptible to windburn because they have the greatest leaf surface area exposed to drying winds. To protect themselves, many broad-leaved evergreens roll their leaves in the winter to reduce the amount of exposed surface area.

Needled evergreens can also burn. If burn has occurred, brown-tipped branches are apparent in the spring when new growth is beginning. As with broad-leaved forms, windburn on conifers is likely to be confined to the outermost branches on the most exposed side of the plant.

Temperature extremes can also cause injury to plants. Plants which are at the limit of their hardiness (termed *marginally hardy*) may be killed by an extended period of severely cold weather. Others may be stunted when all of the previous season's young growth freezes.

After some cold winters, certain plants may show no sign of injury except that their spring flower display is absent. This results if the plant produces its flowers and its leaves in separate buds. The weather may not be cold enough to affect the leaf buds, but freezes the more tender flower buds. This is especially common with forsythia and certain spireas in the northern states.

Unusually warm weather during late winter can also cause plant damage. Fruit trees may be encouraged to bloom prematurely, only to have the flowers killed by a late frost. As a result, the harvest of fruit can be greatly reduced or even eliminated. Spring flowering bulbs can also be disfigured if forced into bloom by warm weather that is followed by freezing winds and snow.

Sun scald is a special type of temperature-related injury. It occurs when extended periods of warm winter sunshine thaw the aboveground portion of a plant. The period of warmth is too brief to thaw the root system, however, so it remains frozen in the ground, unable to take up water. Aboveground, the thawed plant parts require water, which the roots are unable to provide. Consequently, the tissue dries out and a scald condition results.

Sun scald is especially troublesome on evergreens planted on the south side of a building. It also occurs on newly transplanted young trees in a similar location. The young, thin bark scalds easily and the natural moisture content of the tissue is low because of the reduced root system.

Heaving affects turf grass, hardy bulbs, and other perennials when the ground freezes and thaws repeatedly because of winter temperature fluctuations. The heaving exposes the plants' roots to the drying winter wind, which can kill the plants.

Ice and snow damage can occur repeatedly during the winter. The sheer weight of snow and/or ice on plant limbs and twigs can cause breakage and result in permanent destruction of the plant's natural shape, figures 27-2 and 27-3. Evergreens are most easily damaged because they hold heavy snow more readily than deciduous plants. Snow or ice falling off a pitched roof can split foundation plants in seconds.

Plants which freeze before the snow settles on them are even more likely to be injured. Freezing reduces plant flexibility and causes weighted twigs to snap rather than bend under added weight.

Fig. 27-1 Rolled and discolored leaves show the effect of windburn on this rhododendron.

Unfortunately, the older and larger a plant is, the greater is the damage resulting from heavy snow falls and ice storms. There are numerous recorded accounts of the street trees of entire cities being destroyed by a severe winter storm.

Animal damage to plants results from small animals feeding on the tender twigs and bark of plants, especially shrubs. Bulbs are also susceptible. Entire floral displays can be destroyed by the winter feeding of small rodents. Shrubs can be distorted and stunted by removal of all young growth. In cases where the plant becomes *girdled* (with the bark around the main stem completely

Fig. 27-2 Evergreen trees can be broken and even suffer permanent damage from a heavy snow.

removed), the plant is unable to take up nutrients. The plant eventually dies.

Human-Induced Injuries

Certain types of injury are created by people during wintertime landscape maintenance. Some types of injury are due to carelessness on the part of grounds keepers. Other types are the predictable result of poor landscape design. A large number are injuries created because the landscape elements are hidden beneath piles of snow.

Salt injury harms trees, shrubs, bulbs, lawns, and paving. Often the damage does not appear until long after the winter season passes. Thus, the cause of the injury may go undiagnosed.

Fig. 27-3 The weight of ice on the branches of deciduous plants can break and misshape them.

The salt used to rid walks, streets, and steps of slippery ice becomes dissolved in the water it creates. The saline solution flows off walks and into nearby lawns or planting beds. Paving sometimes crumbles under heavy salting. Poured concrete is especially sensitive to this treatment, figure 27-4.

Salt is toxic to nearly all plant life. The resultant injury appears as strips of sterile, barren ground paralleling walks, figure 27-5. Injury can also be seen on the lower branches of evergreens

Snow plow damage can occur to plant and construction materials for several reasons. A careless plow operator may push snow onto a planting or a bench. This often results when those unfamiliar with the landscape are hired to do the snow plowing. Other injuries from plowing are the result of design errors. Plants, outdoor furniture, and light fixtures should not be placed near walks, parking areas, or streets where they will interfere with winter snow removal.

Damage to lawns can result when the plow misses the walk and actually plows the grass, scraping and gouging the lawn, figure 27-6. The grass may not survive if this occurs repeatedly.

Rutting of lawns is the result of heavy vehicles parking on softened ground. When the surface layer of the soil thaws but the subsoil remains frozen, surface water is unable to soak in. Users of the landscape accustomed to finding the ground firm may be unaware of damage caused by vehicles temporarily parked on soft lawns. The soil becomes badly compacted, resulting in unsightly ruts.

REDUCING WINTER INJURY

Some types of winter damage can be eliminated by properly winterizing the landscape in the preceding autumn. Other types can be reduced by better initial designing of the grounds. Still other winter injuries can only be minimized, never totally eliminated.

Windburn can be eliminated in the design stage of the landscape if the planner selects deciduous plant materials rather than evergreen

Fig. 27-4 Excess salt had a damaging effect on these concrete steps. This type of damage is unnecessary.

Fig. 27-5 Dead grass edging the walk in the summer is a symptom of excessive winter salting.

Fig. 27-6 A scraped lawn is the result of a snowplow which has missed the unmarked sidewalk.

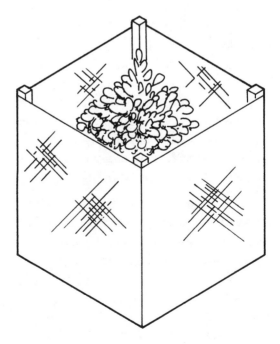

Fig. 27-7 A broad-leaved evergreen protected against windburn with a burlap shield.

materials. The use of deciduous shrubs on especially windy corners can avoid this problem in the garden. If evergreens are important to the design or already exist in the garden, windburn can be reduced by erecting burlap shields around shrubs, figure 27-7. The use of an antitranspirant may also reduce water loss from plants and thereby reduce the effect of windburn. The antitranspirant must be applied in the autumn and again in late winter. While antitranspirants are fairly expensive, they are more practical for the protection of large evergreens than burlap shields.

Temperature extremes can be only partially guarded against. Where wind chill (lowering of temperature because of the force of wind) is a factor, plants should be located in a protected area. Wrapping the plant in burlap also helps. This has proven to be an effective technique for protecting tender flower buds on otherwise hardy plants.

Certain plants, such as roses, can be cut back in the fall and their crowns mulched heavily to assure insulation against the effects of winter. Likewise, any plant that can be damaged by freezing and thawing of the soil should be heavily mulched after the ground has frozen to insulate against premature thawing.

If the landscaper is trying to prolong the life of annual flowers in the autumn, and a frost is forecast, the foliage can be sprinkled with water prior to nightfall. This helps to avoid damage caused by a light frost. The water gives off enough warmth to keep the plant tissue from freezing.

Sun scald of young transplants lessens as the plants grow older and form thicker bark. It can be avoided on trees during the first winter of growth by wrapping the trunks of the trees with paper or burlap stripping. (See Unit 12.) For other types of sun scald, such as that which affects broad-leaved evergreens, the same remedies practiced for windburn and temperature extremes are effective. Wrapping the plants in burlap or the use of antitranspirants gives protection.

Some sun scald can be avoided by the designer. Vulnerable species of plants should not be placed on the south side of a build-

Fig. 27-8 Hinged wooden A-frames protect foundation plantings from damage caused by sliding ice and snow.

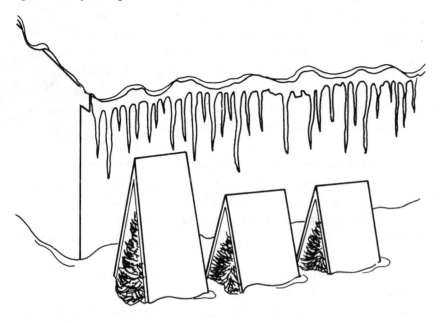

ing, nor should they be placed against a reflective white wall that will magnify the sun's effect on the aboveground plant parts.

Heaving of the turf is impossible to prevent completely. The best defense against it is the encouragement of deep rooting through proper maintenance and good landscaping practices during the growing season.

Bulbs, ground covers, and other perennials can be protected from heaving through application of a mulch after the ground has frozen. The mulch acts to insulate the soil against surface thawing.

Ice and snow damage to foundation plants can be avoided if the designer is careful to not place plants beneath the overhanging roof line of buildings. If the grounds keeper must deal with plants already existing in a danger area, the use of hinged, wooden A-frames over the plants can help to protect them, figure 27-8. As large pieces of frozen snow and ice tumble off the roof, the frame breaks them apart before they can damage the plants.

To aid plants that have been split or bent because of heavy snow accumulation, the grounds keeper must work quickly and cautiously. A broom can be used to shake the snow off the weighted branches. However, the snow removal must be done gently and immediately after the snow stops. If the branches are frozen or the snow has become hard and icy, removal efforts will cause more damage than benefit.

If breakage of plants occurs during the winter, the grounds keeper should prune the damaged parts as soon as possible. This prevents further damage to the plants during the rest of the winter.

Certain plants such as upright arborvitae and upright yews become more susceptible to heavy snow and ice injury as they mature. Often, the damage cannot be repaired. Large and valuable plants in a landscape can be winterized in the autumn by tying them loosely with strips of burlap or twine. (Do not use wire.) When prepared in this manner, the branches cannot be forced apart by the heavy snows of winter, and splitting is avoided.

Animal damage can be prevented by either eliminating the animals or protecting the plants from their feeding. While rats, mice, moles, and voles are generally regarded as offensive, plantings are damaged as much or more by deer, rabbits, chipmunks, and other gentler kinds of animals. Certain *rodenticides* (substances which poison rodents) may be employed against some of the undesirable animals that threaten the landscape. In situations where the animals are welcome but their winter feeding damage is not, a protective enclosure of fine mesh wire fencing around the plants helps to discourage animals from feeding there, figure 27-9.

Salt injury to plants and paving need not be as bad as was illustrated earlier if caution is exercised by the grounds keeper. Salt mixed with coarse sand does a better job than either material used separately. The sand provides traction on icy walks and a small amount of salt can melt a large amount of ice. Excessive salt has no value; it only kills plants and destroys paving. In very cold temperatures, salt does not melt ice; therefore, in these cases, it serves no purpose. Salt can turn compacted snow, which is comparatively safe for traffic, into inches of slush which is messy and even more slippery. The problem of salt injury can be solved by reducing the amount of salt spread on walks and streets during the winter season. The addition of sand distributes the salt more evenly and provides grit for better traction.

Snow plow damage is to be expected if a designer places plants too close to walks and roadways. Therefore, one obvious solution to the problem begins with the designer. When planning landscapes for areas in which winter is normally accompanied by a great deal of snow, the designer should avoid placing shrubs near intersections or other places where snow is likely to be pushed.

Another type of plow damage is the result of the plow driver's inability to see objects beneath the snow. If possible, all objects such as outdoor furniture and lights should be removed from plow areas prior to the winter season. If it is not possible to move them, low objects should be marked with tall, colored poles that can be seen above the snow.

Whenever possible, snow blowers should be used instead of plows. These machines are much less likely to cause damage.

Rutting of lawns usually results from the practice of permitting individuals to park cars on lawns. The best solution to the problem is to avoid doing so. Otherwise, sawhorses or other barriers offer a temporary solution.

Fig. 27-9 A wire enclosure placed around young plants during the winter helps to protect them from animal damage. Even smaller mesh is needed to discourage rodents.

ACHIEVEMENT REVIEW

A. Indicate whether the following types of winter injury are caused by natural conditions (N) or human error (H).

1. crumbled paving resulting from too much salt

2. dry, dead twigs on the windward side of a pine tree

3. dried, blistered bark on the trunk of a recently transplanted tree

4. deep ruts in the lawn in front of a house

5. dead branches in a shrub following an unusually cold winter

6. failure of a shrub to flower in the spring

7. perennials lying on the surface of the soil in the early spring with roots exposed to the drying air

8. dead grass next to a walk heavily salted during the winter

9. an upright evergreen split in the center by snow sliding off a roof

10. a young tree girdled at the base

B. Of the types of winter damage in the following list, which ones could be reduced or prevented by proper winterization of the landscape during the late autumn?

1. crumbled paving

2. sun scald on new transplants

3. breakage of outdoor furniture by snow plows

4. rutting of the lawn by automobiles

5. foundation plants broken by falling snow

6. sun scald on broad-leaved evergreens

7. tree limbs broken off by an ice storm

8. bulbs heaved to the surface of the soil

9. windburn on evergreens

10. flooded basement caused by melting snow

SUGGESTED ACTIVITIES

1. Look for signs of winter injury in nearby landscapes. Find windy corners where evergreens are planted and check for windburned tips. Visit a shopping center, campus, or park where salt is used on walks and parking lots. Look at the paving and nearby plantings for signs of damage. Note the placement of plants in relation to walk intersections and other places where snow may be piled.

2. Conduct ice melting tests. Freeze four pie plates of water. Apply four different mixtures of salt and/or sand to the surfaces and see which melts first. In the first pan, use all salt; in the second, half salt and half sand; in the third, one-quarter salt and three-quarters sand; and in the fourth, all sand. What conclusions can be drawn from the trials? *Note:* Returning the treated pans of ice to the freezer (approximately 20° F) will assure that no natural melting occurs. Check the pans every 15 minutes for observations.

3. Demonstrate the damaging effect of salt upon plant life. Grow some experimental plants in advance. Root each plant in a separate container. Apply only water to some of the plants for a week. To others, apply varying dilutions of a salt and water solution. To a third group, apply water to the soil, but mist the foliage with saltwater several times each day. Record observations of each treatment daily.

GLOSSARY

Accent plant – A plant which is more distinctive than many plants, but does not attract the eye as much as a specimen plant.

Adobe – A heavy soil common to the southwestern United States.

Aeration – The addition of air into the soil; it is accomplished during soil conditioning with materials such as sand or peat moss. It can be encouraged in established lawns by the use of machines called aerators.

Aesthetic – Attractive to the human senses.

Alkaline – Characterized by a high pH.

Angle – The relationship between two joined straight lines.

Annual – A plant which completes its life cycle in one growing season.

Antitranspirant (*also* **antidessicant**) – A liquid sprayed on plants to reduce water loss, transplant shock, windburn, and sun scald.

Arid – A term used in the description of landscapes where there is little usable water.

Balled and burlapped – A form of plant preparation in which a large part of the root system is retained in a soil ball. The ball is wrapped in burlap to facilitate handling during sale and transplanting.

Bare root – A form of plant preparation in which all soil is removed from the root system. The plant is lightweight and easier to handle during sale and transplanting.

Bedding plant – An herbaceous plant preseeded and growing in a peat pot or packet container.

Bulb – A flowering perennial which survives the winter as a dormant fleshy storage structure.

Calibration – The adjustment of a piece of equipment so that it distributes a given material at the rate desired.

Caliche – A highly alkaline soil common to the southwestern United States.

Canopy — The collective term for the foliage of a tree.

Compaction — A condition of soil in which all air has been driven out of the pore spaces. Water is unable to move into and through the soil.

Compass — A graphic design tool used for the construction of circles.

Complete fertilizer — A fertilizer containing nitrogen, phosphorus, and potassium, the three nutrients used in the largest quantities by plants.

Conditioning — Preparation of soil to make it suitable for planting.

Containerized — A form of plant preparation for sale and transplanting. When purchased, the plant is growing with its root system intact within a plastic, metal, or tar paper container.

Contour interval — The vertical distance between contour lines.

Contour lines — Broken lines found on a topographic map. They represent vertical elevation.

Cool-season grass — A type of grass which grows best in temperate regions and during the cooler spring and fall months.

Cost estimate — An itemized listing of the expenses in an operation. It can be applied to a single task or a total project.

Crotch — The point on a tree at which two branches or a branch and the trunk meet.

Crown — The point at which aboveground plant parts and the root system meet.

Cut — A grading practice that removes earth from a slope.

Deciduous — A type of plant which loses its leaves each autumn.

Degree — The unit of measurement for angles.

Diameter — The distance across a circle as measured through the exact center.

Diazo machine — A duplicating machine which makes positive copies from vellum tracings onto heavy paper.

Dormancy (adj.: dormant) — A period of rest which perennial plants experience during the winter season. They continue to live, but have little or no growth.

Drainage — The act of water passing through the root area of soil. Soil is well drained if water disappears in 10 minutes or less from a shrub or tree planting.

Drainage tile — A plastic or clay tube buried beneath the soil which collects excess water from the soil and carries it away.

Enrichment — A contribution made to the outdoor room by a landscape item that is not an element of a wall, ceiling, or floor.

Erosion — The wearing away of the soil caused by water or wind.

Espalier — A form of pruning in which plants are trained flat against a fence or wall. The effect is vinelike and two dimensional.

Evergreen — A type of plant which retains its foliage during the winter. There are needled forms (such as pine, spruce, hemlock, and fir) and broad-leaved forms (such as rhododendron, pieris, euonymus, and holly).

Exotic plant — A plant which has been introduced to an area by human beings, not nature.

Fauna — Animal life.

Fertilization — The addition of nutrients to the soil through application of natural or synthesized products called *fertilizers*.

Fertilizer analysis — The percentage of various nutrients in a fertilizer product. A minimum of three numbers on the fertilizer package indicates the percentage of total nitrogen (N), available phosphoric acid (P_2O_5), and watersoluble potash (K_2O), in that order.

Fill — A grading practice that adds earth to a slope.

Flora — Plant life.

Flower bed — A free-standing planting made entirely of flowers with no background of shrub foliage.

Flower border — A flower planting used in front of a planting of shrubs. The shrubs provide green background for the blossoms.

Focal point — A point of visual attraction. A focal point can be created by color, movement, shape, size, or other characteristics.

Foliage texture — The effect created by the combination of leaf size, sunlight, and shadow patterns on a plant.

Foundation planting — The planting next to a building which helps it blend more comfortably into the surrounding landscape.

Girdling — The complete removal of a strip of bark around the main stem of a plant. After girdling, the ability of nutrients to pass from roots to leaves is lost, causing the eventual death of the plant.

Grading — Changing the form of the land.

Graft — A man-made bond between two different plants, one selected for its aboveground qualities (scion) and the other for its below ground qualities (stock).

Ground cover — A low-growing, spreading plant, usually 18 inches or less in height.

Grounds keeper — A professional engaged full-time in landscape maintenance.

Hardiness — The ability of a plant to survive through the winter season.

Heading back — A pruning technique which shortens a shrub branch without totally removing it.

Heaving — An action which causes shallowly rooted plants, such as grasses, ground covers, and bulbs, to be forced to the surface of the soil. The action results from repeated freezing and thawing of the soil surface.

Herbaceous — A type of plant which is nonwoody. It has no bark.

Herbicide — A chemical used to kill weeds.

Holdfasts — Special appendages of certain vines that allow them to climb.

Incurve — The center of a corner planting bed and a natural focal point.

Inert material — Filler material which has no purpose other than to carry and dilute active ingredients in a mixture.

Inorganic — Consisting of nonliving materials.

Intangible — A quality denoting something that cannot be touched.

Jump-cut — A pruning technique for the removal of large limbs from trees without stripping bark from the trunk. It involves a series of three cuts.

Landscape architect — A licensed professional who practices landscape planning, usually on a scale larger than residential properties.

Landscape contractor — A professional who carries out the installation of landscapes.

Landscape designer — A professional who devotes all or part of a work day to the design of landscapes.

Landscape installation — The actual construction of the landscape.

Landscape maintenance — The care and upkeep of the landscape after installation.

Landscape nurseryman or woman — A professional who is concerned with the sale and installation of landscape plants and related materials.

Landscaping — A profession involving the design, installation, and maintenance of the outdoor human living environment.

Lateral bud — Any bud below the terminal bud on a twig.

Leaching — The dissolving of materials (such as nutrients) in the water which is present in soil, causing the material to quickly pass the point at which plant roots can benefit from them.

Lime — A powdered material used to correct excess acidity in soil.

Loam — Soil which contains approximately equal amounts of clay, silt, and sand (a desirable condition).

Mulch — A material placed on top of soil to aid in water retention, prevent soil temperature fluctuations, or discourage weed growth.

Native plant — A plant which evolved naturally within a certain locale.

Naturalized plant — A plant which was introduced to an area as an exotic plant, but which has adapted so well that it may appear to be native.

Noxious weeds — Persistent weeds defined by law in most states. They are perennial and difficult to control. The presence of these weeds in a grass seed mix indicates that the seed is of low quality.

Nutrient ratio — A comparison of the proportion of each nutrient in a fertilizer to the other nutrients in the same fertilizer. Example: A fertilizer with a 5-10-5 analysis has a 1-2-1 ratio of ingredients.

Organic — Consisting of modified plant or animal materials.

Outcurve — The sides of a corner planting.

Perennial — A plant which lives more than two growing seasons. It usually is dormant during the winter.

Pesticide — A chemical used for the control of insects, plant diseases, or weeds.

pH — A measure of the acidity or alkalinity of soil. A pH of 7.0 is considered neutral. Ratings below 7.0 are acidic, above 7.0 alkaline (basic).

Plant list — An alphabetical listing of the botanical names of plants used in a landscape plan, their common names, and the total number used.

Plugging — A method of lawn installation which uses cores of live, growing grass.

Propagation — The reproduction of plants. It may be sexual or asexual (by vegetative cuttings, layering, etc.).

Protractor — A graphic design tool for measuring angles.

Pruning — The removal of a portion of a plant for better shape or more fruitful growth.

Puddling — Compaction of soil to such a degree that water will not soak into it.

Purity — The percentage, by weight, of the pure grass seed in a mixture.

Quickly available fertilizer — A fast-action fertilizer which has its nitrogen in a water soluble form for immediate release into the soil.

Radius — One-half of the diameter of a circle.

Riser — The elevating portion of a step.

Scaffold branch — A lateral branch of a tree.

Scale (engineer's) — A measuring tool which divides the inch into units ranging from 10 to 60 parts.

Serif — A decorative stroke attached to a letter to create an ornate appearance.

Shrub — A multistemmed plant smaller in size than a tree.

Silhouette — The outline of an object viewed as dark against a light background.

Site — An area of land having potential for development.

Slope — A measurement that compares the horizontal length to the vertical rise or fall of land. The measurement can be determined from a topographic map.

Slow-release fertilizer — A slow-action fertilizer in which the nitrogen content is in a form not soluble in water. The nitrogen is released more slowly into the soil for more efficient intake by plants.

Sodding – A method of lawn installation which uses strips of live, growing grass. It produces an immediate effect on the landscape, but is more costly than seeding.

Soil texture – The composition of a soil as determined by the proportion of sand, silt, and clay which it contains.

Species – A category of plant classification distinguishing the plant from all others.

Specimen plant – A plant that is highly distinctive because of such qualities as flower or fruit color, branching pattern, or distinctive foliage. Its use creates a strong focal point in a landscape.

Spreader – A garden tool used for the even distribution of materials such as grass seed and fertilizer.

Stolon – An underground stem which can cause the propagation and spreading of certain shrubs.

Sucker – A succulent branch which originates from the root system. The vegetation of suckers is abnormal and undesirable.

Sun scald – A temperature-induced form of winter injury. The winter sun thaws the aboveground plant tissue, causing it to lose water. The roots remain frozen, and thereby unable to replace the water. The result is drying of the tissue.

Symbols – Drawings which represent overhead views of trees, shrubs, or other features of a landscape plan.

Tangible – A quality denoting something that is touchable.

Tender – A condition of plants which implies their lack of tolerance to cold weather.

Tendrils – Special appendages of certain vines that allow them to climb.

Terminal bud – The end bud on a twig.

Terrain – The rise and fall of the land.

Thatch – Dead, semidecomposed grass clippings on the surface of soil.

Thinning out – A pruning technique which removes a shrub branch at or near the crown of the plant.

Topiary – A form of pruning in which plants are severely sheared into unnatural shapes such as animals or chess pieces.

Topography – A record of an area's terrain.

Trace elements – Nutrients essential to the growth of many plants, but needed in far less amounts than the major elements.

Transplant – To relocate a plant.

Tread – That portion of a step on which the foot is placed.

Triangle – A three-sided graphic design tool. It commonly has either a 30°-60°-90° or a 45°-90°-45° combination of angles.

T-square – A long straightedge which takes its name from its shape. It is a graphic design tool.

Twining – One method by which certain vines are able to climb.

Unit cost – The price of the smallest available form of an item described in a cost estimate.

Vellum – A thin, paperlike material on which a landscape plan is traced.

Warm-season grass – A grass which grows best in warmer regions of the country and during the summer months.

Water sprout – A succulent branch which grows from the trunk of a tree. The vegetation of water sprouts is abnormal and undesirable.

Weed – A plant growing where it is not wanted and having no economic value.

Weep hole – A means of preventing water buildup behind a retaining wall.

Windburn – Drying out of plant tissue (especially evergreens) by the winter wind.

Winter injury – Any damage done to elements of the landscape during the cold weather season of the year.

Working drawing – A copy of a landscape design done on heavy paper or plastic film. The working drawing is used repeatedly during actual construction of the landscape and must be very durable.

Wound paint – A sealing paint used over plant wounds of 1 inch or more in diameter after pruning.

ACKNOWLEDGMENTS

Technical Reviewer: Paul Blair, Board of Cooperative Educational Services, Albany-Schoharie-Schenectady Counties, New York

Consulting Editor, Agricultural Series: H. Edward Reiley, Frederick County Board of Education, Frederick, Maryland

The Staff at Delmar Publishers

Goldberg and Rodler Landscape Contractors, Huntington, NY, Color Plates

Millard Irwin

Anthony Markey, figures 4-3, 4-9, 11-6, 12-6, 12-7, 14-2, 15-1, 15-2, 15-4, 22-2, 22-3, and 22-6

Alan R. Nason, figure 1-5

Stanley Pendrak, figure 2-4

Walter Ressler, figures 1-7, 4-8, 6-1, and 11-4

J.S. Staedtler Co., Inc., figures 4-1, 4-5, and 4-7

State University of New York at Cobleskill, figure 19-3

Joseph Tardi Associates

Harold Toles, figures 2-2 and 2-5

United States Department of Agriculture, figures 12-1, 14-1, 15-5, 17-6, 21-2, 21-3, and 22-8

Ray Wyatt, figures 16-2, 16-3, 16-4, and 16-6

The remaining photographs are the original work of the author.

Illustrator: Johnny Orozco

The author wishes to express appreciation to the following for their help in the preparation of this text:

Stewart Allen, Director, The Allen Organization

E. Mark Barry

Lothar Baumann

Robert C. Bigler Associates, Architects

Michael Boice, Designer

Martin Bozak

Richard Centolella

Cooperative Extension Service, Cornell University Ithaca, New York

Joann Cornish, L.A.

Bernard Cushman

Vickie Davis

Edward Dennehy, Designer

Russell Ireland, Landscape Contractor

David Johnson

John Krieg, L.A.

Peter Lee

Anne Lovell

Mark Magnone, L.A.

Anthony Markey, L.A.

Joye Noth

Stanley Pendrak

Daniel Pierro, Designer

Margaret Porter, Designer

Olga Ressler

Robert Rodler, Landscape Contractor

Robin Schutte

Frederick Smith

Ricky Sowell

Paul Stacey

State University of New York Agricultural & Technical College, Cobleskill, NY

Kathleen and Richard Vedder

Unit 24, Pricing Landscape Maintenance, was based on an original chart by David Lofgren which appeared in *Grounds Maintenance Magazine,* January 1968.

Landscaping: Principles and Practices was classroom tested at the State University of New York Agricultural and Technical College at Cobleskill, New York.

INDEX